T0332527

GEOLOGY OF JAPAN

DEVELOPMENTS IN EARTH AND PLANETARY SCIENCES

GEOLOGY OF JAPAN

Edited by

Mitsuo HASHIMOTO

Dept. of Earth Sciences, Ibaraki University, Mito, Japan

Developments in Earth and Planetary Sciences

08

Terra Scientific Publishing Company/Tokyo

Kluwer Academic Publishers
Dordrecht/Boston/London

Library of Congress Cataloging-in-Publication Data

CIP DATA APPEARS ON SEPARATE CARD

ISBN 0-7923-0909-X (Kluwer)
ISBN 4-88704-106-3 (Terrapub)

Published by Terra Scientific Publishing Company (TERRAPUB),
302, 303 Jiyugaoka Komatsu Building, 24-17 Midorigaoka 2-chome, Meguro-ku,
Tokyo 152, Japan,
in co-publication with Kluwer Academic Publishers, Dordrecht, Holland

Sold and Distributed in the U.S.A. and Canada
by Kluwer Academic Publishers,
101 Philip Drive, Assinippi Park, Norwell, MA 02061, U.S.A.
in Japan by Terra Scientific Publishing Company (TERRAPUB),
302, 303 Jiyugaoka Komatsu Building, 24-17 Midorigaoka 2-chome, Meguro-ku,
Tokyo 152, Japan

In all other countries, sold and distributed
by Kluwer Academic Publishers,
P.O. Box 322, 3300 AH Dordrecht, Holland

(This book is published by Grant-in-Aid publication of Scientific Research Result
of the Ministry of Education, Science and Culture of Japan)

Printed in Japan

PREFACE

This book is an English translation of "Geology of Japan" published as the 15th volume of the Earth Sciences Series in 1980 from Iwanami Shoten, Tokyo. In preparing this English edition, however, the content has been considerably abridged and re-written, in order to make it suitable for non-Japanese readers. Consequently, this is an abridged and revised English version of the original Japanese edition.

There is no unique way to describe the geology of an area, however, three alternative methods are often adopted. The first method describes the large scale geological construction of each rock terrane and mass; its structure, rock types, etc. A second method describes the geology of an area in term of stratigraphical succession, for example, from Paleozoic to Cenozoic systems. A third approach places the emphasis mainly on the historical development of the region. Each descriptive method is, of course, not completely separable from another. However, the first method will be adopted here. The principal elements characterising the geology of the Japanese islands will be described for each geotectonically defined terrane and mass. The last chapter will also include an outline of the geohistorical development of Japan.

The authors of the original book were: M. Hashimoto (Chaps. 1, 2 and 3), S. Ishihara (Chap. 4), K. Kanehira (Chap.9), K. Kanmera (Chaps. 2 and 10), T. Matsuda (Chaps. 5 and 10), Y. Naruse (Chaps. 6 and 7), N. Niitsuma (Chap. 8), T. Sato (Chap. 8), and T. Ui (Chap. 7), and the editorial work was done by M. Hashimoto, K. Kanmera and T. Matsuda.

Acknowledgements

The translator wishes to express his sincere thanks to the co-writers of the original version for their close cooperation, to Iwanami Shoten for having given their permission for the translation of the original manuscript, to Ms. V. Emery for revising the translated manuscript and to Mr. K. Oshida for his arrangement and help. Finally, we regret very much that Professor Keiichiro Kanehira passed away in October 1985, and also, that we were not able to have his valuable suggestions for this English version.

Mitsuo Hashimoto

CONTENTS

CHAPTER 1

INTRODUCTION

The Japanese Islands constitute part of the long garland-like chain of islands which fringes the northwestern margin of the Pacific Ocean, extending from the Aleutians through Japan to the Philippines. In considering the configuration of islands, together with the submarine topography, it can be seen that Japan consists of two trench-arc systems: the East Japan and the West Japan Island Arcs. The East Japan Island Arc marks the boundary between the Eurasian, the Pacific and the Philippine Sea plates, and the West Japan Island Arc marks the boundary of the Eurasian and the Philippine Sea plates. Recent studies suggest, however, that the westernmost tip of the North American plate may extend to Northeast Japan, so that a part of Northeast Japan may lie on the North American plate.

The geology of the Japanese Islands is so extremely complicated that it is not easy to obtain an overall view even from small-scale geological maps such as the 1:1,000,000 scale issued by the Geological Survey of Japan (HIROKAWA *et al.*, 1978). However, when we try to form a generalised picture, it seems reasonable to consider the geology of the Japanese Islands to be comprised of the following five elements (Figs. 1.1(a), (b), (c) and (d)):

    1)   Pre-Neogene sedimentary and regional metamorphic rocks.

    2)   Granites and rhyolites mostly of Late Mesozoic to Early Tertiary age.

    3)   Neogene sediments and associated volcanics.

    4)   Quaternary sediments.

    5)   Plio-Pleistocene volcanics.

1.1   *Pre-Neogene Sedimentary and Regional Metamorphic Rocks*

Ages of rocks dealt with under this heading range in age from Silurian to Paleogene. These rocks are observed at various locations in the Japanese Islands. The substantial part of the Pre-Neogene rocks are of Triassic to Jurassic age. Rocks older than Late Carboniferous are fairly restricted in

3

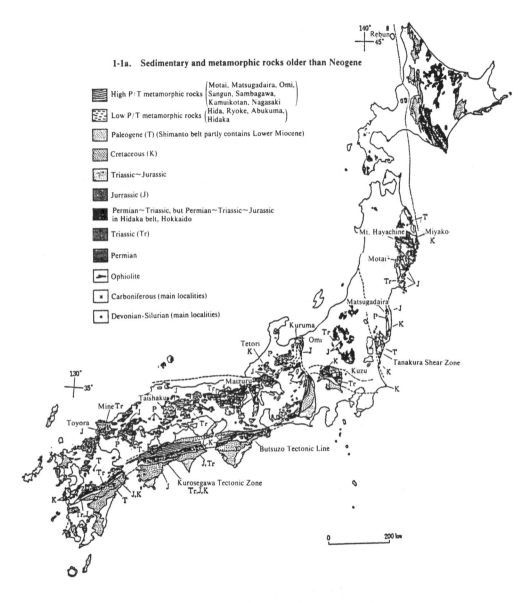

1-1a.   Sedimentary and metamorphic rocks older than Neogene

High P/T metamorphic rocks (Motai, Matsugadaira, Omi, Sangun, Sambagawa, Kamuikotan, Nagasaki)

Low P/T metamorphic rocks (Hida, Ryoke, Abukuma, Hidaka)

Paleogene (T) (Shimanto belt partly contains Lower Miocene)

Cretaceous (K)

Triassic~Jurassic

Jurrassic (J)

Permian~Triassic, but Permian~Triassic~Jurassic in Hidaka belt, Hokkaido

Triassic (Tr)

Permian

Ophiolite

x   Carboniferous (main localities)

•   Devonian-Silurian (main localities)

1-1b. Granites and rhyolites

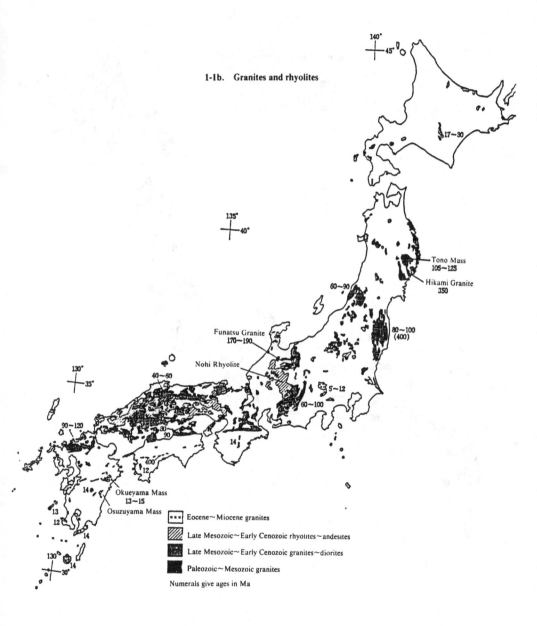

Tono Mass
105~125

Hikami Granite
350

Funatsu Granite
170~190

Nohi Rhyolite

Okueyama Mass
13~15
Osuzuyama Mass

┄ Eocene~Miocene granites

▨ Late Mesozoic~Early Cenozoic rhyolites~andesites

▥ Late Mesozoic~Early Cenozoic granites~diorites

■ Paleozoic~Mesozoic granites

Numerals give ages in Ma

1-1c. Sedimentary and volcanic rocks of Neogene

Neogene sedimentary rocks
Pliocene~Early Pleistocene volcanic rocks
Miocene volcanic rocks

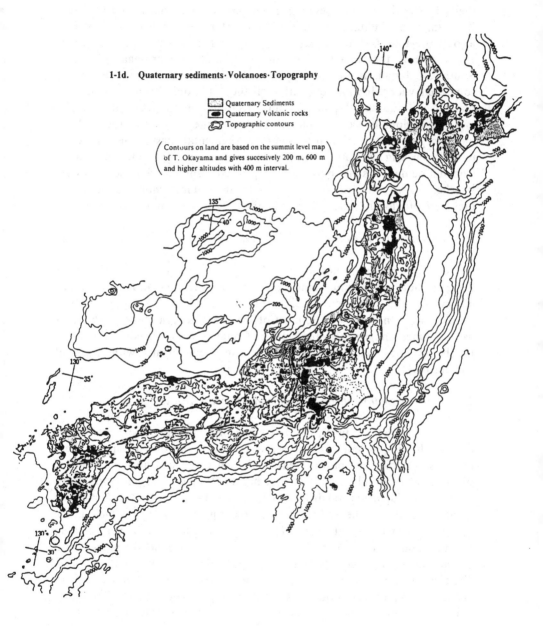

1-1d.   Quaternary sediments·Volcanoes·Topography

Quaternary Sediments
Quaternary Volcanic rocks
Topographic contours

Contours on land are based on the summit level map
of T. Okayama and gives succesively 200 m, 600 m
and higher altitudes with 400 m interval.

their distribution, and they are only exposed regionally in the South Kitakami and Abukuma regions of Northeast Japan, while tectonic blocks, also of pre-Late Carboniferous age are sporadically distributed in narrow zones, such as the Hida Border Fault Zone and the Kurosegawa Zone of Southwest Japan. Rocks of Upper Carboniferous to Permian age show somewhat wider distribution, although they also only occur in limestone plateaus in Southwest Japan, and in the tectonic zones mentioned above.

The regionally metamorphosed rocks are the main constituent of the metamorphic belts, some also occur as fragmental masses and blocks in tectonic zones, in association with serpentinite. Incidentally, bedded cupriferous iron-sulfide deposits and poorly developed sedimentary manganese ores are interbedded with the pre-Neogene sedimentary and metamorphic rocks. The ores were previously widely worked, although most are now abandoned.

1.2   Granites and Rhyolites, Mostly of Late Mesozoic to Early Tertiary Age

Granites and rhyolites are widely exposed, particularly in the Inner Zone of Southwest Japan. Their radiometric ages range mainly between 120 and 60 Ma, although some are either older (about 180 Ma) or younger (less than 30 Ma). In other words, most of the Japanese granites and rhyolites were intruded and/or extruded between Cretaceous and Paleogene times. A number of tungsten and molybdenum deposits are associated with these acid igneous rocks.

1.3   Neogene Sediments and Associated Volcanic Rocks

Late Cenozoic sediments widely cover pre-Paleogene rocks mentioned above. This is particularly the case in central to northeast Japan. These deposits are thick in coastal areas, and each geological horizon continue offshore, while in inland regions they become thinner. Neogene formations in the Sea of Japan coastal district of Northeast Japan are the product of one great cycle of marine sedimentation, the upper layers of which grade gradually into the Quaternary. The lower Neogene is also very rich in volcanic and pyroclastic rocks, which have been frequently altered into greenish propylitic rocks. A number of Kuroko deposits occur in these rocks. The term "Green Tuff" used to describe this type of greenish altered Tertiary volcanic rock, has been widely used among Japanese geologists. The upper part of the Neogene formations occasionally include petroleum and natural gas.

1.4 Quaternary Sediments

Japan is a very mountainous country, with only less than one-sixth of its total area comprising the Japanese "plain". The "plain" comprises the lowland plain together with the terraced uplands of a slightly higher altitude. The large cities of Japan have all been developed on the plain, where more than 70% of Japanese people dwell. The plain is underlain by Quaternary sediments, which are often intercalated by thin beds of volcanic ash, mostly derived from nearby volcanoes.

1.5 Plio-Pleistocene Volcanoes

About 200 volcanoes, some of which are still active, occur in the Japanese Archipelago. The volcanoes have usually retained their original shape, and volcanoes form an important feature of the Japanese natural landscape, and almost three quarters of the national parks in Japan are located in the volcanic districts. Geothermal areas associated with the young volcanoes are widely also developed as hot-spring resorts, and are also utilized for energy.

CHAPTER 2

PRE-NEOGENE SEDIMENTARY AND METAMORPHIC ROCKS

This chapter is concerned with pre-Neogene sedimentary strata and the regional metamorphic rocks derived from them. They range from Silurian to Paleogene in age. A substantial part of these rocks are younger than Permian, and pre-Carboniferous rocks are fairly restricted in distribution generally occurring in narrow tectonic zones, such as the Hida Marginal and the Kurosegawa Zones. However, in the South Kitakami and Abukuma regions, they are comparatively widely exposed (SAITO and HASHIMOTO, 1982).

The pre-Neogene geology of the Japanese islands has been classified into a number of terranes, which are usually separated from each other by faults or tectonic zones. Figure 2.1 shows the currently accepted terrane classification of Japanese pre-Neogene geology (HIROKAWA et al., 1978). In Southwest Japan in particular, a conspicuous zonal arrangement of tectonically stretched terranes has been observed, and generally speaking, the rocks become younger southwards from the Sea of Japan to the Pacific side. The terranes of Southwest Japan will be described first, commencing from north (older) to south (younger), and next the terranes of Northeast Japan and Hokkaido will be described.

I. SOUTHWEST JAPAN

2.1 Hida Metamorphic Terrane

The Hida metamorphic terrane occupies the north central part of Honshu, the main island of Japan. The terrane appears to extend westwards to Oki, a small island off Matsue in the Sea of Japan. The eastern and southern margins of the Hida terrane are limited by a narrow arcuate tectonic zone called the Hida Marginal (or Border) Zone, while the northern and western parts are covered by Tertiary to Quaternary sediments. Within

Fig. 2.1.   Classification of the old rock terranes (slightly modified after Hirokawa *et al.*, 1978).
1: Hida, 2: Hida Border Zone, 3: Sangun, Chugoku and Tamba-Mino, 4: Maizuru, 5: Ashio, 6:
Ryoke, 7: Sambagawa, 8: Chichibu and Sambosan, 9: Shimanto, 10: Abukuma, 11: South
Kitakami, 12: North Kitakami, 13: Ishikari-Kamuikotan, 14: Hidaka-Tokoro.

the Hida terrane, the Kuruma (Lower Jurassic) and the Tetori (Upper
Jurassic to Lower Cretaceous) Groups partly cover the metamorphic rocks,
in the northeastern and central to western districts, respectively. Furthermore, the Upper Cretaceous rhyolites and granites were extruded and
intruded into the south-central and northeastern parts of the terrane (Fig.
2.2).

Metamorphic rocks of the Hida terrane consist mainly of medium- to
coarse-grained biotite-gneisses, hornblende-gneisses, calc-silicate gneisses

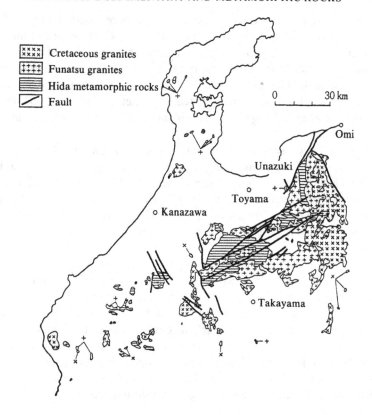

Fig. 2.2.   Metamorphic and granitic rocks of the Hida terrane (after HIROI, 1978).

and marble. The biotite-gneisses occasionally contain pyralspite garnet, sillimanite, andalusite, cordierite, and occasionally corundum and spinel. The calc-silicate gneisses are characterized by diopside, wollastonite, grandite, scapolite and chondrodite. When the original carbonate rocks were dolomitic, the metamorphic derivatives contain forsterite and phlogopite. Detailed studies of the mineral parageneses have not yet been worked out for the Hida metamorphic area; however, sporadic occurrences of andalusite and cordierite, in the biotite-gneisses, indicate that the Hida terrane belongs to the andalusite-sillimanite type facies series. Recently, rocks considered to be indicative of the granulite facies have been reported from Oki Island and from the main part of the Hida terrane (SUZUKI, 1973).

The Hida metamorphic rocks were widely intruded by the Funatsu Granite and have experienced contact metamorphism as a result. Radiometric age of the granite has been determined as 180 Ma, while gneisses that remained unaffected by contact metamorphism were dated as 240 Ma.* It is therefore considered that the principal part of the Hida metamorphic rocks were formed during Permo-Triassic times and that intrusion of the Funatsu Granite took place later in Jurassic times. Furthermore, some metamorphic rocks show ages older than 300 Ma. Certain geologists consider that the Hida metamorphic rocks were repeatedly metamorphosed, the earliest metamorphic event having taken place in late Precambrian to early Paleozoic time.

Crystalline schists containing staurolite and kyanite, known as the Unazuki Schists, have been reported from the northeastern part of the Hida terrane. They form a separate geological unit belonging to the kyanite-sillimanite type facies series (HIROI, 1981). The schists were in part thermally metamorphosed by the Funatsu Granite. Radiometric age dating of unaffected schists, however, gives 210–250 Ma, and this is comparable to the age of the Hida gneisses. Rocks similar in lithology are found in a zone surrounding the main part of the Hida terrane along its eastern and southern margins.

The Kuruma Group consists mainly of interstratified sandstone and shale as well as a number of coarse-grained sandstone beds, and conglomerate occur at both the base and top of the Group. The lithofacies changes from non-marine in the lower part, through marine in the middle part to non-marine again in the upper formations, forming a clearly defined sedimentary cycle. In the Tetori Group, on the other hand, three cycles of sedimentation have been observed. The middle cycle of the Tetori Group is a brackish to freshwater facies yielding abundant fossil plants, including large tree trunks.

2.2 Hida Marginal Zone

As stated in the previous section, the Hida metamorphic terrane includes the Unazuki Schist Belt, which is surrounded by a tectonic zone called the Hida Marginal Zone. This is an arcuate narrow belt in which several tectonic blocks and masses, comprising various rock types are tectonically mixed up. Rocks exposed in this zone include: sedimentary rocks of Silurian to Permian age, conglomerates of unidentified age, crystalline schists and amphibolites, together with serpentinite and related rocks.

Silurian to Permian strata consist mostly of thick limestone beds, together with acidic tuffs of Lower Devonian age. Lower Carboniferous and Lower Permian strata mostly consist of greenstones. The limestones yield abundant fossils, these include Devonian corals, brachiopods and trilobites,

---

*Radiometric age data are mostly cited from NOZAWA (1977)

Carboniferous fusulinids and algae, and Permian fusulinids. Crystalline schists of the Hida Marginal Zone are mostly the product of high P/T metamorphism, however occasionally rocks of low P/T type are also found. Serpentinite pervasively intrudes between the boundaries of different rock types and is frequently associated with meta-gabbros and leucocratic rocks such as rodingite, jadeitite and albitite. Radiometric ages of metamorphic rocks do not indicate a single event, but are classified into the following three groups: approximately 680 Ma, 440 Ma and 370–300 Ma.

The geological features of the Hida Marginal Zone described above indicate that it is a serpentinite tectonic melange.

2.3 Sangun Metamorphic Belt

The Sangun metamorphic belt does not seem to form a belt, but consists of a number of fragmented and separated areas underlain by low-grade metamorphic rocks. They are randomly distributed in the Chugoku and northern Kyushu districts.

Metamorphic rocks of this belt are mostly derived from pelitic to psammitic sedimentary rocks, and basic to intermediate volcanic and volcaniclastic rocks, and a few associated rocks are derived from cherts and limestones. The geological structure is somewhat complicated. Strata of the original rocks appear to form gently folded structures, whose axes run roughly parallel to an E–W direction. A number of serpentinite or serpentinized peridotite masses of various sizes are accompanied by metamorphosed gabbros which were intruded into the Sangun belt and the Maizuru Zone to the south.

The Sangun metamorphic belt is a glaucophanitic terrane, and contains characteristic minerals such as blue amphiboles, pumpellyite and stilpnomelane which occur widely in rocks of basic composition, although lawsonite is rare in this belt.

Four metamorphic zones are classified in the Sangun belt. They are, from low to high grade (Fig. 2.3; HASHIMOTO, 1972):

1) Pumpellyite-chlorite zone,
2) Pumpellyite-actinolite zone,
3) Epidote-glaucophane zone, and
4) Barroisite zone.

The petrological features of these zones are similar to those of the Sambagawa metamorphic belt described below (See p. 29).

The peridotite masses are distributed in two parallel rows (Fig. 2.3). Peridotite masses in the southern row have a layered structure, while those of the northern row are massive. In addition to the commonly observed serpentinization, contact metamorphism due to later granitic intrusions have

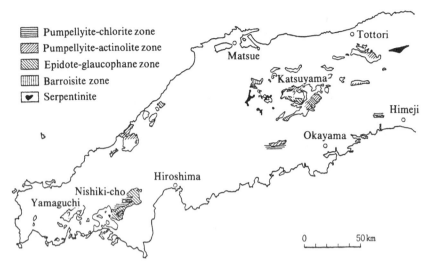

Fig. 2.3.   Facies of the Sangun metamorphic terrane (after HASHIMOTO, 1972).

also modified the mineralogy of the peridotites. However detailed studies on the massive Tari-Misaka bodies, for example, suggest that they may have originally consisted of dunite, harzburgite and chromitite. The innermost part of the contact aureole is characterized by crystallization of Ca-poor orthopyroxene and olivine as thermal metamorphic minerals. On the other hand, the original rock types of the layered peridotite masses are dunite, lherzolite, wehrlite and clinopyroxenite. Tremolite is a contact metamorphic mineral of the layered masses.

Stratigraphical and paleontological data, though scarce, suggest that the original rocks of the Sangun metamorphic belt are largely Permo-Carboniferous in age. On the other hand, radiometric dating gives at least two groups of metamorphic ages. The older is 300–270 Ma and the younger 180–170 Ma, other subordinate data show ages of about 260 Ma and 210 Ma (NISHIMURA et al., 1983). The older ages are in conflict with the results of stratigraphic and paleontologic studies, indicating that the metamorphic age has not been definitively worked out yet.

2.4   Paleozoic and Mesozoic Sedimentary Rocks of the Chugoku, Tamba and Mino Terranes

The Upper Paleozoic strata are widely exposed in scattered localities of the Chugoku terrane which extends from northern Kyushu to eastern

Chugoku (Fig. 2.4). The Chugoku strata are classified into three lithofacial groups.

The first lithofacies comprises large masses of reef-limestone complex, underlain by basic volcanic bodies (KANMERA and NISHI, 1983). Detailed studies of fusulinid biostratigraphy were made on limestone plateaus, exemplified by that of the Akiyoshi district, western Chugoku, where 21 biozones were established. According to these studies, the age of the limestone masses range from Early Carboniferous (Viséan) to Middle Permian. Sedimentary formations, of non-calcareous but varied lithofacies, are distributed around the limestone plateaus. They consist of siliceous tuff, spiculite, sandstone, mudstone and conglomerate and many limestone blocks. The non-calcareous formations, may represent either talus deposits derived from reef-limestone complexes or olistostromes.

The second and third lithofacies of the Upper Paleozoic formations in the Chugoku district largely consist of mudstone and sandstone. However, the second lithofacies group is associated with abundant basic volcanics and the third lithofacies with plentiful chert. Most of the rocks would seem to be of Middle to Upper Permian age, but a few of the third lithofacies group are of Triassic to Jurassic age.

In the Triassic formations of the Chugoku district only two lithofacial groups are observed. The first lithofacies group (Ladinian to Norian) consists of shallow marine to non-marine sediments, and the second lithofacies group (Scythian to Norian) are fully marine. The two groups are separately distributed from each other, and their geological and stratigraphical relations have not yet been ascertained. The shallow marine to non-marine Triassic sediments overlie either the Sangun metamorphic rocks or the Paleozoic formations mentioned above. Sedimentary cycles commencing with conglomerate and coarse sandstone and closing with fine sandstone and mudstone are observed in the Triassic formations, and a few coal seams are intercalated in the middle part of the cycle (Fig. 2.5; TOKUYAMA, 1958). The second lithofacies is fully marine and consists principally of very thick Triassic mudstone and chert deposits. Recent studies of fossil conodonts, included in the chert, suggests the existence of complex imbricate structures formed from a number of thrusted sheets (TOYOHARA, 1976). All the Triassic conodont biozones, ranging from Scythian to Norian age, have been recognized. Even one bedded chert pile of only 60 m in thickness contains all the biozones. Furthermore, several blocks of various sizes comprising sandstone, greenstone and limestone containing Permo-Carboniferous fusulinids, were enveloped in mudstone of the marine lithofacies formations, suggesting they are olistostromal.

Jurassic sedimentary rocks underlie a few areas of the Chugoku district. Abundant ammonites and plant fossils have been found in a section of the

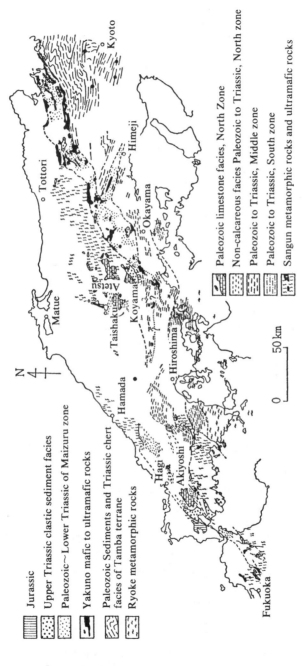

Fig. 2.4.   Classification of the older rocks in the Chugoku to Western Kinki provinces (compiled after HASE, 1964; TAMBA BELT RESEARCH GROUP, 1975; HIROKAWA *et al.*, 1978).

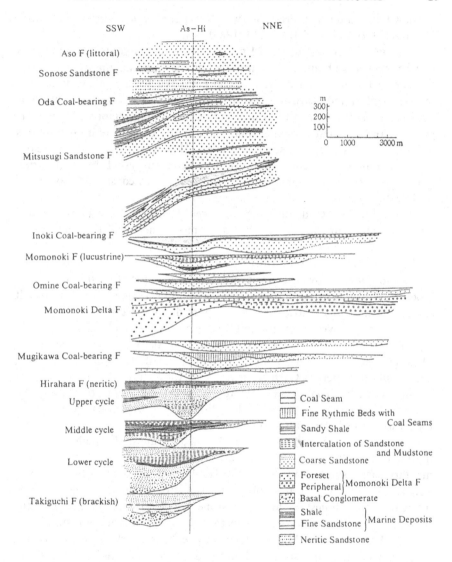

Fig. 2.5. Idealized lithostratigraphic succession of the Upper Triassic Mine Group (after TOKUYAMA, 1958). As-Mi :Asou-Hirahara line. F: Formation.

Jurassic rocks, indicating that they were formed during shallow sea to bay environments. The formations are in fault contact with, or overlie the Sangun metamorphic rocks.

Paleozoic to Mesozoic sedimentary rocks, of similar lithofacies to those of the Chugoku terrane, are more widely exposed in the Tamba and Mino terranes (MIZUTANI and HATTORI, 1983).

In the central part of the Mino terrane, a few large limestone masses overlie greenstone bodies, as exemplified by the Akasaka limestone plateau. Similar to the Akiyoshi district in the Chugoku terrane, these limestone masses yield abundant fusulinid and algal fossils of Permo-Carboniferous age. The greater part of the Mino terrane, other than regions of limestone plateaus and associated rocks, is underlain by thick sedimentary formations of Triassic mudstone, sandstone and chert. However, in this terrane also a highly complex imbricate structure has been suggested by radiolarian biostratigraphy, and upper horizons of some thrusted sheets may be of Middle Jurassic to Early Cretaceous age.

On the other hand, limestone is rarely found in the Tamba terrane, which is underlain largely by layers of mudstone, sandstone and chert intercalated with a few greenstone beds (Fig. 2.6) (SAKAGUCHI, 1974). Fossil evidence indicates that the principal part of the strata is of Permo-Triassic age, although the Lower Triassic has not been found. Olistostromes with blocks of Permo-Carboniferous limestone, middle Triassic chert and basalt embedded in muddy matrices are also reported in the Tamba terrane.

Finally, it is worth noting that Jurassic radiolarians have been widely reported from bedded cherts and mudstones of the Chugoku, Tamba and Mino terranes. The strata consisting of the bedded cherts and mudstones, and containing permo-Carboniferous limestone masses were previously considered to be of late Paleozoic age, however, the recent results suggest that the strata are olistostromal or accretionary, and were constructed during Late Jurassic or Early Cretaceous time.

The Ashio (Fig. 2.7), Asahi and Yamizo terranes, although located to the far east in the Kanto district, are underlain by Paleo-Mesozoic sedimentary rocks, similar to those of the Tamba and Mino terranes mentioned above. They largely consist of chert, sandstone and shale with a few intercalations of limestone and greenstone. The Nabeyama limestone in the Ashio terrane rests on massive greenstone and yields abundant fusulinids of the late Early to early Middle Permian age. It is overlain by the Norian Adoyama Group, consisting of limestone, green shale and chert with an angular unconformity, which represents the largest Paleozoic-Mesozoic stratigraphic gap observed in Japan. The Adoyama Group is further covered by a sandstone bed, which seems to be of Jurassic age.

2.5   Cretaceous Rocks of the Chugoku and Tamba Terranes

Non-marine Cretaceous sediments are developed in limited areas of the Chugoku and Tamba terranes. They rest unconformably on older sedi-

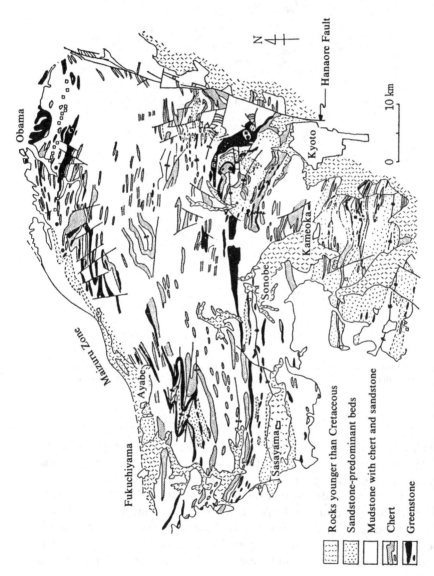

Fig. 2.6. Outline of the distribution of older rocks in the Tamba area (after TAMBA BELT RESEARCH GROUP, 1975). Fossil ages ○: Lower Early Permian, □: Upper Early Permian, ●: Lower Middle Permian, ■: Middle Middle Permian (these four based on fusulinids), +: Middle to Late Triassic (based on conodonts).

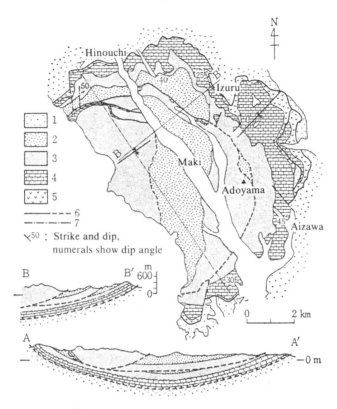

Fig. 2.7. Geologic map and cross-section of the Kuzuu district in the Ashio terrane (after YANAGIMOTO, 1973). 1 and 2: Jurassic, 3: Triassic chert, 4 and 5: Permian, 6: Thrust, 7: high-angle fault.

mentary and metamorphic rocks. The lowest Cretaceous is represented by the Toyonishi Group, which is in turn overlain by the Lower Cretaceous Kanmon Group. The upper part of the Kanmon Group contains large amount of andesitic to dacitic volcanics (MURAKAMI, 1975). Volcanic rocks became predominant in the Upper Cretaceous, and are associated with intrusive rocks of similar compositions, forming a few volcano-plutonic associations in the Late Mesozoic to Early Cenozoic in these regions. The acid to intermediate volcanic and plutonic rocks of this age are described in Chapter 4.

2.6   Maizuru Zone

The Maizuru Zone comprises a tectonic belt, 15–20 km in width and about 180 km in length, which it extends in an ENE-WSW direction from the Maizuru district to the eastern part of the Chugoku province. The principal constituents of the Maizuru Zone are the Middle to Upper Permian Maizuru Group, the Lower to Middle Triassic Yakuno Group and the Yakuno basic igneous rocks. The Yakuno basic igneous rocks are exposed in two separated narrow belts running along the northern and southern margins of the Maizuru Zone, and the Maizuru Group and the Yakuno Group are exposed in the intervening region.

The Maizuru Group is divided stratigraphically into a lower and an upper part. The lower part mainly consists of slate and greenstone, while the upper part consists of intercalated shale and sandstone. The Maizuru Group contains bivalves, brachiopods and fusulinids of Upper Permian age.

The Yakuno Group overlies the Maizuru Group above an angular unconformity, it comprises a lower and an upper part. In the lower part, three heterotopic sedimentary facies predominated, these are respectively firstly conglomerate and coarse sandstone, secondly medium to fine sandstone, and thirdly shale. These have been developed side-by-side from northwest to southeast. The thickness of the beds decreases southeastwards, and faunal variation of fossil bivalves is also in keeping with the changes in facies and thickness. These features indicate that the sedimentary environments experienced during deposition of the lower part of the Yakuno Group would have changed steadily southeastwards, from deltaic through neritic to bathyal (Fig. 2.8). On the other hand, the upper part of the Yakuno Group largely consists of fine-grained sediments, containing fossil cephalopods, and does not show any significant facies or faunal changes observed in the lower part.

The upper Triassic (Carnian to Norian) Nabae Group are sporadically distributed in the Maizuru Zone, above the Lower to Middle Triassic Yakuno Group with local angular unconformities. The sediments are neritic in origin and contain abundant bivalves.

The Yakuno basic igneous rocks comprise a variety of rock types such as basalts and their associated pyroclastics, gabbros and peridotites, although the rocks have all been metamorphosed. Detailed geological and petrological studies indicate that the basic igneous rocks and their associated sedimentary rocks constitute an ophiolite complex. It is comprised of the following five parts in descending order:

1)   Basalts and hyaloclastic together with clastic sediments,
2)   Gabbros,

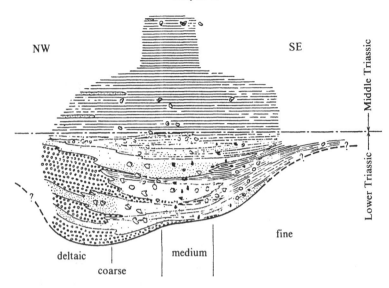

NW   SE

Middle Triassic

Lower Triassic

deltaic

coarse

medium

fine

Fig. 2.8.   Idealized litho- and bio-stratigraphic succession of the Lower to Middle Triassic in the Maizuru Zone (after NAKAZAWA, 1958).

3)   Pyroxenites,
4)   Dunites, and
5)   Harzburgites.

Metamorphic grade changes progressively from the prehnite-pumpellyite facies in 1) through the amphibolite facies in 3) to the granulite facies in 4). (Fig. 2.9).

2.7   Ryoke Metamorphic Belt

To the south of the Chugoku, Tamba and Mino terranes lies the Ryoke metamorphic belt. From the Shiojiri-Takato district, central Honshu, it extends southwestwards through the central part of the Kinki province and into the Setouchi region to the eastern part of Kyushu. The Ryoke metamorphic belt extends for about 700 km in length and 30–50 km in width. Its southern margin is limited by the Median Tectonic Line which separates it from the Sambagawa belt to the south, but in the western half of the contact area, Ryoke belt rocks are covered by the Upper Cretaceous Izumi Group and do not appear to be in direct contact with the Sambagawa rocks. The northern boundary of the Ryoke belt, on the other hand, is not clearly defined, the lowest grade metamorphic rocks seem to grade imperceptively into Paleozoic-Mesozoic sedimentary rocks of the Tamba and Mino terranes,

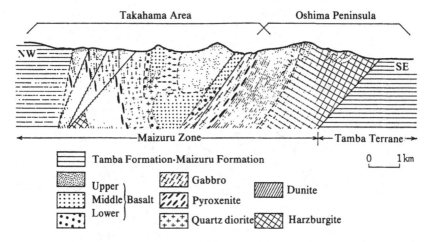

Fig. 2.9.   Idealized cross-section of the Yakuno ophiolite complex (simplified after ISHIWATARI, 1978).

although fault relations between them have been claimed in some localities.

In the Ryoke metamorphic belt, granitic rocks predominate over metamorphic rocks, and the latter are fragmentarily exposed in a few areas such as the Shiojiri-Takato, Shidara-Hazu, Kasagi and Iwakuni-Yanai districts. Metamorphic rocks of the Ryoke belt are mostly derivatives of pelitic, psammitic and cherty rocks, while derivatives of limestone and basic volcanic rocks are rare. Fossil evidence is insufficient to indicate the age of the original sediments, however, recent studies on conodont and radiolarian remains show that they are largely Upper Triassic to Middle Jurassic.

The Ryoke metamorphic belt is of the andalusite-sillimanite type facies series. Metamorphic zoning has been made on the basis of mineralogical variation of pelitic rocks. The grade generally increases from north to south. ONO (1977) classified the following three zones in the Shiojiri-Takato district:

1) Chlorite-biotite zone,
2) Biotite zone, and
3) Sillimanite zone (Fig. 2.10).

Chlorite disappears at the boundary between zones 1) and 2), and sillimanite and cordierite occur in zone 3). Andalusite is found in the lower grade part of the sillimanite zone. The association of potassium feldspar with such minerals as andalusite and cordierite has been widely observed and is characteristic of the Ryoke belt. Recent studies have revealed that staurolite is rather common in pelitic assemblages in the Hazu district (ASAMI and HOSHINO, 1980).

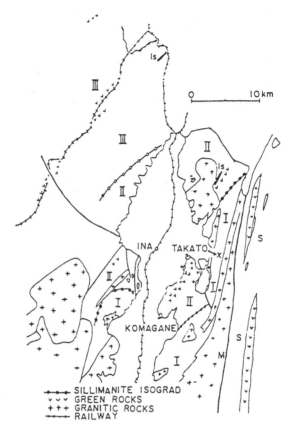

Fig. 2.10.   Mineral zones of the Ryoke metamorphic terrane, Takato district (after ONO, 1969).
I: Sillimanite zone, II: Biotite zone, III: Chlorite-biotite zone, ls: Limestone, M: Median
Tectonic Line, S: Sambagawa terrane.

Abundant granitic rocks were intruded into the metamorphic rocks and
altered them. The contact aureoles show crystallization of porphyroblastic
muscovite and replacement of sillimanite by andalusite and muscovite. The
granitic rocks were grouped into a number of separate intrusive masses, on
the basis of their geological field relations and petrographical features, and
into those intruded before and those intruded after the extrusion of Nohi
Rhyolites. However, the radiometric ages for most of granitic rocks show a
narrow age range of 60–90 Ma. Furthermore, radiometric data of the
metamorphic rocks also shows similar values. On the other hand, the
following geological observations indicate that the Ryoke low P/T regional
metamorphism would have taken place somewhat earlier than the extrusion

of Nohi Rhyolites and the intrusion of the younger granites.

    1)   The configuration of regional metamorphic zoning is not conformable with the distribution of granitic rocks,

    2)   The granitic intrusions induced contact metamorphism in the regional metamorphic rocks, and

    3)   The geological structures of the regional metamorphic rocks are fairly complicated, although the Nohi Rhyolites have not been so strongly deformed.

2.8   Izumi Belt

On the northern side of the Median Tectonic Line, particularly on its western half, runs a narrow belt consisting of thick piles of intercalated sandstone and shale together with a few thin beds of acidic tuff. The belt extends for about 300 km with a maximum width of about 12 km. The sediment piles are collectively called the Izumi Group, and are of late Cretaceous age. The strata generally dip 30°–50° to the east and form part of a synclinal structure. The northern wing of the Izumi belt lies unconformably over rocks of the Ryoke belt, and the southern wing is in fault contact with the Sambagawa schists. A large part of the Izumi Group is marine, yielding fossil shells such as *Inoceramus*. The basal conglomerate of the northern wing contains pebbles of granite, quartz porphyry and mica schist which were all derived from the Ryoke belt to the north (Fig. 2.11).

Thick piles of sediment similar to those of the Izumi Group are also exposed in Kyushu (MATSUMOTO, 1954). These form the Onogawa Group in the eastern part and the Mifune and Goshora-Himeura Groups in the western part of the island. Eleven cycles of sedimentation have been recognized in the Onogawa Group (Turonian to Santonian), but lateral lithofacies variations are conspicuous. The Mifune Group (Cenomanian to Turonian) comprises one sedimentary cycle. This ranges from the lowermost non-marine sediments, through the lower brackish to shallow marine sandstone-siltstones to the upper varicolored siltstone-shales. The Goshora Group (Albian to Turonian) consists mostly of coarse marine sediments but a few non-marine red beds occur in both the basal and uppermost horizons. The Himeura Group lies above a slight unconformity (Coniacian to Maastrichtian), which are also composed of shallow marine coarse-grained sediments.

2.9   Sambagawa Metamorphic Belt

The Sambagawa belt is the largest regional metamorphic terrane in Japan. It starts in the Kanto Mountains, central Honshu, extends westwards

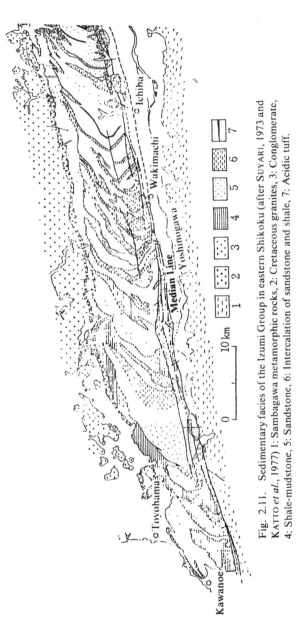

Fig. 2.11.   Sedimentary facies of the Izumi Group in eastern Shikoku (after SUYARI, 1973 and
KATTO et al., 1977) 1: Sambagawa metamorphic rocks, 2: Cretaceous granites, 3: Conglomerate,
4: Shale-mudstone, 5: Sandstone, 6: Intercalation of sandstone and shale, 7: Acidic tuff.

through the Kami-Ina and Tenryu districts, crosses the Kii Peninsula and Shikoku island from east to west and reaches the eastern part of Kyushu. Its total length is about 1,000 km. The Sambagawa belt pinches and swells from district to district, and its maximum width of about 35 km is observed in central Shikoku, where it comprises an area of crystalline schists, Mikabu Green Rocks and a part of the Chichibu belt to the south.

Detailed geological studies have been carried out, particularly in Shikoku, where the Sambagawa belt has been divided into sub-belts, from north to south:

1) The belt of typical crystalline schists,
2) The southern marginal zone,
3) The belt of Mikabu Green Rocks, and
4) The northern part of the Chichibu belt.

The Mikabu Green Rock sub-belt largely consists of weakly metamorphosed basic volcanic and pyroclastic rocks. It is traceable intermittently, and relatively large masses are exposed in the Okuki, Osugi and Sanagochi areas of Shikoku. Outside of Shikoku similar masses have been, for example, developed in the Toba district of Kii peninsula, the Tonmakuyama area of Shizuoka Prefecture and Mount Mikabu in the Kanto province. The complex derives its name from Mount Mikabu. The second and the fourth sub-belts are in direct contact with each other in some places, where an anticline has been observed. A fault, called the Kamiyakawa-Ikegawa Line, lies at the axis of the anticline, along which small masses of serpentinite or andesite of Tertiary age have been intruded.

The crystalline schist formations in central Shikoku are classified stratigraphically as follows (KOJIMA et al., 1956).

	Upper Sub-group	Ojoin Formation
		Minawa Formation
	Middle Sub-group	
Yoshinogawa Group		Koboke Formation
		Kawaguchi Formation
	Lower Sub-group	
		Oboke Formation

The largest is the Minawa formation which mainly consists of thick piles of green schist beds, together with some quartz schists. Within the upper part of this formation a few masses of meta-gabbro and meta-peridotite are embedded, the former being exemplified by the Irazu mass and the latter the Higashiakaishi mass. Furthermore, most of the bedded cupriferous iron-sulfide ore deposits are intercalated in the Minawa Formation, however, they have now been largely worked out. Crystalline schists, other than those of the Minawa Formation are mostly pelitic and psammitic along with a few quartzose rocks.

Geologic structures have also been studied in central Shikoku, and two large-scale recumbent folds and one fold with a subvertical axial plane have been recognized (HARA *et al.*, 1977). One of the recumbent folds is the Nagahama Fold in western Shikoku and it contributes to the formation of a nappe structure named the Nagahama Nappe (Fig. 2.12). Corresponding nappe structures are also developed in other areas of the Sambagawa belt. In central Shikoku however, another nappe, called the Tsuji-Saruta Nappe overlies the Nagahama Nappe, and in its southernmost part, a few minor recumbent folds such as the Shirataki overturned anticline are observed (Fig. 2.13).

Since the work of Seki in 1958 in the Kanto Mountains district, many papers have been published on the metamorphic petrology of the Sambagawa belt, the most extensive studies having been carried out in central Shikoku. (Fig. 2.14) (BANNO, 1964; ERNST *et al.*, 1970; BANNO *et al.*, 1986). According to these studies, the Sambagawa metamorphic belt, including the Mikabu

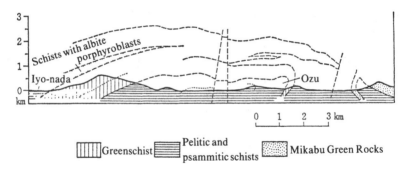

Fig. 2.12.   Cross-section of the Nagahama recumbent fold (after HIDE, 1972; HIDE *et al.*, 1977).

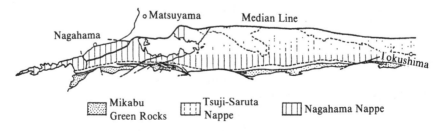

Fig. 2.13.   Tectonic outline of the Sambagawa belt in Shikoku (after HARA *et al.*, 1977).

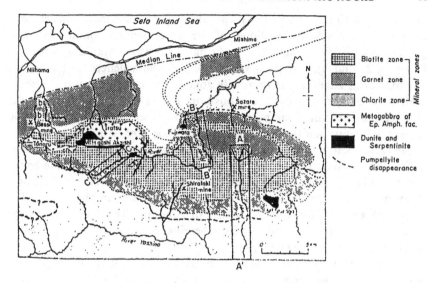

Fig. 2.14.   Mineral zones of the Sambagawa metamorphic terrane in central Shikoku (after BANNO *et al.*, 1986). The mark "b" denotes the area where biotite schists occur but not fully studied. Rectangular outlines inside the figure show the areas studied in detail (see the original text).

Green Rocks and a part of the Chichibu belt, can be divided into the following five zones. From low to high grade, they are:
1) Pumpellyite-chlorite zone,
2) Pumpellyite-actinolite zone,
3) Epidote-glaucophane zone,
4) Barroisite zone, and
5) Zone of the epidote-amphibolite facies.

Generally, 1) includes the northern part of the Chichibu belt, and 2) comprises the Mikabu Green Rocks and the southern marginal zone of the crystalline schist area. The area of the typical crystalline schists comprises zones 3), 4) and 5).

The pumpellyite-chlorite assemblage is widely observed in rocks of basic composition in zone 1). The degree of metamorphic recrystallization is so low that most rocks of this zone retain their original sedimentary and volcanic textures and minerals. Zone 2) is characterized by the wide occurrence of the pumpellyite-actinolite association in basic rocks. Glaucophane and lawsonite are occasionally found. Pelitic rocks of zone 2) show remarkable schistosity. Pumpellyite and lawsonite become unstable in common basic rocks of zone 3), and the only Ca-Al hydrous silicate occurring is epidote. Basic rocks, relatively rich in iron include crossite in association with epidote. Pelitic schists show a notable banding structure in addition to the schistosity. Some of the pelitic schists contain Mn rich garnets. A calciferous amphibole in the basic schists is actinolite, this occurs in zones 2) and 3), but its place is taken by subcalciferous hornblende (barroisite) in zone 4), the barroisite zone. A remarkable feature of these schists, particularly of pelitic composition, in this and higher zones, is the wide development of albite porphyroblasts. Pyralspite garnet is a principal mineral of pelitic schists. Zone 5) represents the highest grade part of the Sambagawa belt. Common hornblende is a characteristic mineral of basic schists in this zone. Biotite appears widely and porphyroblastic plagioclase becomes slightly calcic in pelitic schists in the higher grade part of this zone. A few masses of meta-gabbro, recrystallized into epidote amphibolite, and of peridotite are embedded in the schists of zone 5). The former carry such notable rocks as kyanite- and/or zoisite-bearing amphibolites, quartz eclogite and hornblende eclogite (BANNO et al., 1976).

Among many peridotite masses in the Sambagawa belt, the Higashiakaishi and the Shiragayama masses of central Shikoku, and the Ryumon mass of Kii Peninsula, in particular, have been petrologically studied in detail. The principal rock types of the Higashiakaishi mass are wehrlite, clinopyroxenite, garnet clinopyroxenite (eclogite) and chromitite. Parageneses and element partitions among the associated minerals suggest that the peridotite mass has been equilibrated under amphibolite facies conditions. However, the nature of spinel-bearing clinopyroxenite and granulite found in the meta-gabbros closely associated with the peridotite mass indicate that they may have previously experienced much higher temperature conditions. On the other hand, the Shiragayama and the Ryumon masses consist largely of dunite formed through metamorphic recrystallization of serpentinite.

A gabbro-peridotite complex of the Toba district Mikabu Green Rocks terrane, eastern Kii Peninsula, consists of dunite, wehrlite and plagioclase peridotite in parts and consists of olivine hornblendite and hornblende gabbro in other parts. The peridotites are considered to have been formed as cumulates precipitated from a picrite magma. Most rocks of the complex have been equilibrated under the pumpellyite-actinolite facies conditions, the principal metamorphic minerals being crysotile, hydrogarnet, chlorite and diopside (See p. 60).

Geological evidence indicates that the metamorphism of the Sambagawa belt took place later than Early Cretaceous, as the various radiometric age datings cluster around 90 Ma.

2.10   Chichibu Belt (in a Broad Sense)

To the south of the Sambagawa crystalline schists belt and the Mikabu

Green Rocks terrane, three belts run parallel to each other in an E-W direction. These are the Chichibu belt (in a narrow sense) and the Sambosan belt. Both belts mostly consist of weakly to non-metamorphic rocks, with the Kurosegawa Tectonic Zone inbetween. The three are collectively called the Chichibu Belt in a broad sense. The Chichibu and the Sambosan belts extend from the Kanto Mountains district, central Japan, through the Akaishi Mountains and the Kii Peninsula to Shikoku, and finally reach the eastern part of Kyushu. Their total length is about 1,000 km. On the other hand, the easternmost part of the Kurosegawa Zone lies in the eastern part of the Kii Peninsula; the zone extends westwards through Shikoku to the Yatsushiro district in the western part of Kyushu. Its length is about 600 km.

2.10.1 Chichibu belt (in a narrow sense)

The Chichibu belt is considered to be a subduction complex consisting principally of late Triassic to middle Jurassic mudstone, sandstone and chert. It has a highly complicated imbricate structure consisting of a number of thrusted sheets and lenticular masses. Furthermore, abundant blocks and masses of limestone and greenstone are embedded as olistoliths in sheared mudstones or are bordered by faults. Some limestone masses yield fusulinids and algal fossils indicating Permian age, and others bivalves, gastropods and ammonites indicate Triassic age. A few limestone masses are Carboniferous and contain corals, fusulinids and algae of this age. An edifice of two stacked Permian submarine basaltic volcanoes has been analysed in detail in the Chichibu terrane of eastern Shikoku (MARUYAMA and YAMASAKI, 1978). On the other hand, upper Jurassic to Cretaceous strata both consisting mostly of ordinary clastic sediments are developed in a few basinal areas near to the Kurosegawa Zone to the south.

2.10.2 Sambosan belt

Similarly to the Chichibu belt, the Sambosan belt has a highly complicated imbricate geologic structure and would seem to be a subduction complex. Generally speaking, this belt consists of the following three units:

1)   Sandstone, shale and reef limestone of middle to upper Jurassic age occurring in the northern marginal part of the belt,

2)   Chert, sandstone and shale of middle to upper Triassic age in the central part, which form contemporaneous units heterotopic to the next unit, and

3)   Chert, micritic limestone and basaltic rocks of middle to upper Triassic age occur in the southern part of the belt (KANMERA, 1969).

Along the Kurosegawa Zone to the north, a number of scattered areas underlain by Cretaceous strata ranging from Berriasian to Campanian, are developed. The strata show several cycles of sedimentation, each of which commences with basal conglomerate and sandstone yielding fossil plants and fresh-water to brackish shells, and changes in facies through intercalated sandstone-mudstone, to fine-grained mudstone containing marine bivalves, gastropods, echinoderms and ammonites at the close. The Cretaceous formations as a whole become increasingly richer in marine sediments with ascending stratigraphic horizon.

2.10.3 Kurosegawa Tectonic Zone

The Kurosegawa Tectonic Zone is a fault zone running through the midst of the Chichibu belt (in a narrow sense) (Fig. 2.15; MARUYAMA et al., 1984). The fault zone is represented by a number of lenticular masses of various kinds of rocks intermittently distributed along it. Rocks of the zone are dissimilar to those of the contiguous Chichibu belt and comprise Siluro-Devonian sedimentary and pyroclastic rocks, Permo-Triassic and upper Triassic sediments, high- and low-grade metamorphic rocks, sheared granites, serpentinite and others. The serpentinite intrudes pervasively into massive boundaries of different rock types and plays an important role in the geology of this fault zone. Recent studies of the radiometric ages of various metamorphic rocks show that there are at least four groups: 400 Ma, 394–352 Ma, 327–317 Ma and 240–208 Ma. Consequently, the rocks of the Kurosegawa Tectonic Zone are highly variable in lithology and in age, and they have all been invaded by serpentinite along the boundaries. These geologic features indicate that the fault zone is a serpentinite tectonic melange zone.

The Kurosegawa Zone becomes wider westwards and eventually it is in direct contact with the Sambagawa schist belt in the eastern part of Kyushu, where the Chichibu belt (in a narrow sense) diminishes.

2.11 Shimanto Belt

The Chichibu belt (in a broad sense) thrusts over the Shimanto belt to the south with a low-angle fault called the Butsuzo line.

The Shimanto belt occupies the southernmost part of the zonal structure of Southwest Japan pre-Neogene terranes. It starts at the Kanto Mountains area, central Japan, and extends westwards through the Pacific coast of Honshu, Shikoku and Kyushu. Some geologists claim that the southern extension of the belt can be traced into Okinawa-Honto, the largest island of the Ryukyu archipelago. The Shimanto belt is divided into the northern and the southern sub-belts (Fig. 2.16). The former consists of upper Jurassic to Cretaceous strata, and the latter of Paleogene to lower Miocene. All the strata are collectively called the Shimanto Supergroup. However, post-Late Miocene sediments lie above an angular unconformity, the older rocks of the

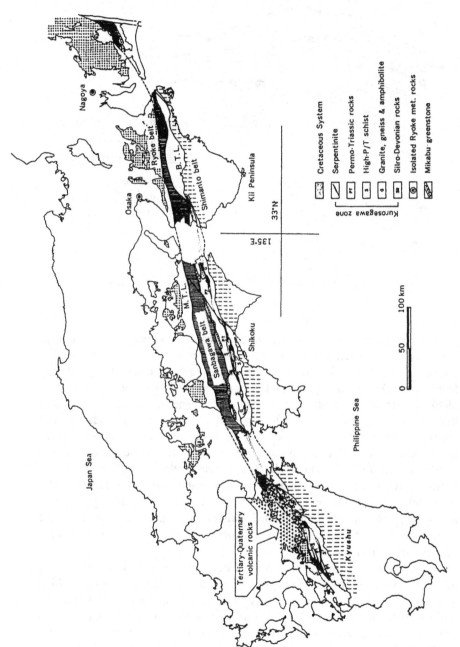

Fig. 2.15. Areal extent in relation to the geology of Southwest Japan of the Kurosegawa zone (after MARUYAMA *et al.*, 1984).

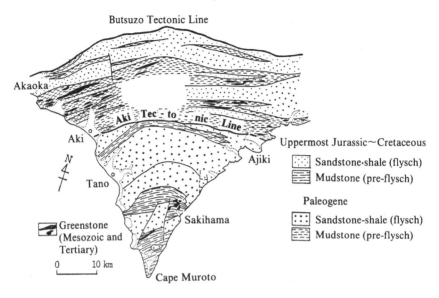

Fig. 2.16.   Geological outline of the Shimanto terrane of the Muroto Peninsula, Shikoku.

southern sub-belt.

The northern sub-belt has an imbricate geologic structure comprised of a number of thrusted sheets, which mostly dip northwards with medium angles. Two contrasting sedimentary facies are recognized among the formations. One consists principally of a mudstone facies in which abundant blocks of greenstone and chert are embedded, and the other consists of a flysch facies and is composed of rhythmical intercalations of sandstone and mudstone along with minor acidic tuff. The former facies predominates in the northern part of this sub-belt, while the latter predominates in the southern part. It is claimed that the flysch facies strata generally occur higher in the stratigraphic horizon than the predominantly mudstone facies strata. A few sets of the two facies types constitute the northward dipping imbricate structures, and contain thrusts which cut the lower part of each mudstone formation, and the strata of the northern sub-belt as a whole become younger from north to south.

The southern sub-belt is also similarly, composed of two contrasting sedimentary facies strata: the mudstone facies and the flysch facies with thick sandstone beds. The former contains many blocks of greenstone and chert, and olistostromal structures are observed in some places. The greenstones are comprised of massive and pillowed basalt flows, volcanic breccias and hyaloclastites, and are often associated with overlying cherts. Sandstone beds of the flysch facies occasionally show graded-bedding and sole marks, which suggest that they are turbidites. The formations of the above two facies constitute a set, with the flysch facies beds overlying the mudstone facies strata. A few sets are in subparallel arrangement along the extension of the Shimanto belt. The ages of the formations generally become younger from north (Eocene) to south (Oligocene to early Miocene).

Sedimentary and acid pyroclastic rocks younger than Middle Miocene unconformably cover the beds of the Shimanto Super-group, and a few masses of Miocene granite have been intruded into them.

II. NORTHEAST JAPAN

2.12 Abukuma Metamorphic Belt

The Abukuma metamorphic belt extends, in a north-south direction, from the Fukushima-Soma area to the Hitachi district. It has an areal extent of about 150 km in length and about 45 km maximum width. The northeastern and the southwestern margins are bordered by fault zones, the latter of which is called the Tanakura Shear Zone. The northwestern and the southeastern parts of the belt are covered by Tertiary and Quaternary sediments.

The Abukuma belt is largely underlain by granitic rocks, but in the southern half regional metamorphic rocks are relatively widely exposed (Fig. 2.17). The metamorphic rocks are classified into two groups: The Gosaisho and the Takanuki metamorphic rocks. The Gosaisho occupy the eastern part of the metamorphic terrane and are predominantly greenschists and fine-grained amphibolites, together with minor mica schists. The Takanuki metamorphic rocks are exposed in the western area and consist largely of mica gneisses associated with subordinate amphibolites and calcic gneisses. The geological relationship of the two groups has not yet been fully worked out due to the highly complex folded and faulted structures. However, some geologists consider that the Gosaisho metamorphics represent an upper horizon compared to the Takanuki metamorphics.

The Abukuma metamorphic belt is a classical area for the andalusite-sillimanite type facies series (MIYASHIRO, 1958, 1961; Shido, 1958). It is divided into three metamorphic mineral zones. They are, from low- to high-grade, Zones A, B and C. Zone A is of the greenschist facies and is characterized by a wide occurrence of the epidote-actinolite-chlorite assemblage in rocks of basic composition. On the other hand, hornblende appears widely distributed in basic rocks of Zones B and C, indicating that both zones are of the amphibolite facies. The transition of andalusite to sillimanite takes place at the boundary between Zones B and C. Furthermore, in Zone C cordierite and potassium feldspar are common in pelitic rocks. It has been

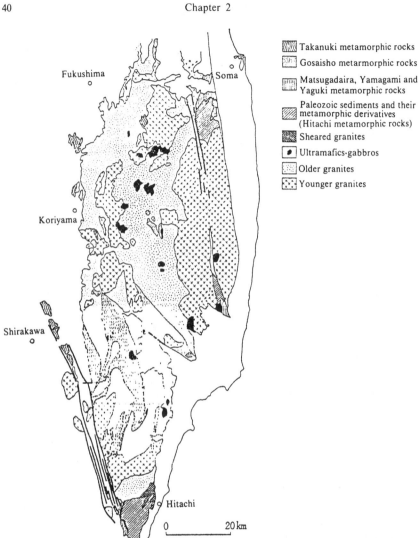

Fukushima

Soma

Koriyama

Shirakawa

Hitachi

0   20 km

	Takanuki metamorphic rocks
	Gosaisho metarmorphic rocks
	Matsugadaira, Yamagami and Yaguki metamorphic rocks
	Paleozoic sediments and their metamorphic derivatives (Hitachi metamorphic rocks)
	Sheared granites
	Ultramafics·gabbros
	Older granites
	Younger granites

Fig. 2.17.   Geological outline of the Abukuma metamorphic terrane (after KANO *et al.*, 1973).

noted that zones of the epidote amphibolite facies are practically absent in this metamorphic belt.

However, in recent studies, kyanite and staurolite, which are considered to be characteristic of the medium-pressure type metamorphism, have been found to occur sporadically but widely. These minerals seem to have relict textural relations under the microscope. Some geologists thus claim that in the Abukuma belt repeated metamorphism would have taken place, the earlier event being of the kyanite-sillimanite type and the later event of the andalusite-sillimanite type (KANO *et al.*, 1973).

Regional metamorphic rocks slightly different from, but essentially similar to, the rocks of the Gosaisho-Takanuki area, are exposed in the Hitachi district, the southernmost part of the belt. Chloritoid and anthophyllite are frequently found, and a zone, if not wide, of the probable epidote amphibolite facies is recognized in this district (TAGIRI, 1973).

Granitic rocks of the Abukuma belt are classified into two groups, the older and the younger granites, on the basis of the field relations. The older granites are harmonious in shape with the surrounding rocks and show conformable internal structures. However the younger granites are usually disharmonious with the country rocks.

Fossil evidence indicating the age of the original sediments is rare, and has hardly been obtained even from the low-grade metamorphic rocks. But Viséan corals and early Permian fusulinids have been reported from low-grade rocks of the Hitachi district. The radiometric dating of the metamorphic rocks and the granitic rocks gives respectively 100 to 110 Ma and 90 to 100 Ma. Little differences are found between data for the older and the younger granites. It would seem that the geological history of the Abukuma metamorphic belt is fairly complicated and cannot be fully interpreted by the existing data.

2.13   South Kitakami Terrane

In the central part of the Kitakami Mountains, Northeast Japan, an arcuate serpentinite belt called the Hayachine-Goyozan belt, runs roughly in a NW-SE direction. It separates the South Kitakami from the North Kitakami terranes. The pre-Neogene geology of the former is quite different in various aspects from the latter (Fig. 2.18).

The geologically defined South Kitakami terrane is considered to be allochthonous to other Japanese pre-Neogene terranes, including the North Kitakami (SAITO and HASHIMOTO, 1982). There the strata of Silurian to Lower Carboniferous are regionally exposed, although rocks of similar ages are quite restricted in their distribution along narrow belt-like areas such as the Hida Marginal Zone and the Kurosegawa Zone. Furthermore, an almost complete stratigraphic column representing this time interval is obtained in this terrane (Fig. 2.19).

The lowest strata comprise the Middle (?) to Upper Silurian Kawauchi Formation, which are the oldest fossiliferous sediments known in Japan (ONUKI, 1956). The Kawauchi Formation mostly consists of limestone along

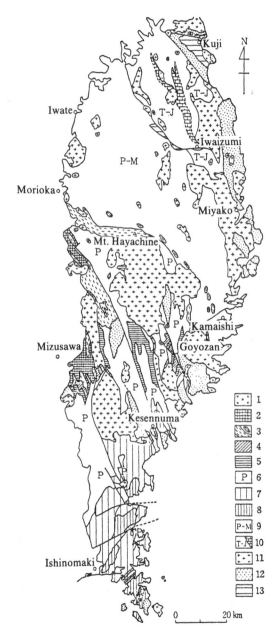

Fig. 2.18.   Geological outline of the kitakami Mountains. 1: Hikami granites, 2: Motai Group,
3: Ultramafic rocks, 4: Siluro-Devonian, 5: Carboniferous, 6: Permian, 7: Triassic, 8 and 9:
Permo-Triassic, 10: Triassic to (?) Jurassic, 11: Early Cretaceous granites, 12: Early Cretaceous
volcanic and sedimentary rocks, 13: Paleogene.

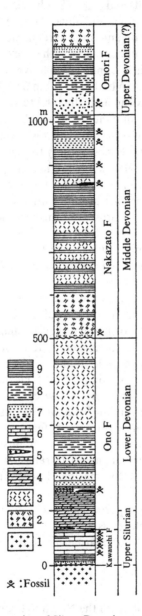

Fig. 2.19.   Idealized columnar section of Siluro-Devonian strata in the South Kitakami area.
1: Hikami granites, 2: Basic lavas and pyroclastics, 3: Intermediate to acidic lavas and
pyroclastics, 4: Pale green to reddish purple siliceous sediments, 5: Boulder breccia, 6:
Limestone, 7: Conglomerate and sandstone, 8: Intercalations of sandstone and shale, 9: Shale,
F: Formation.

with subordinate acidic tuff and shale. The limestone yields abundant fossils such as corals, stromatoporoids and algae. The Devonian deposits overlying the Silurian are rich in volcanic and pyroclastic rocks of intermediate to acidic composition. The upper part of the Devonian is characterized by bedded limestone, which is dissimilar to massive facies of the same age in the limestone plateaus, resting on bodies of submarine basaltic rocks distributed in the Chugoku, Tamba, Mino and Chichibu terranes in Southwest Japan. A part of the Upper Devonian contains fossil plants such as *Lepidodendron*. The Devonian grades upwards into the Lower Carboniferous without any discontinuity, and the latter has essentially the same lithology as the former. Some of the pelitic rocks of this age are lateritic, and rich in alumina and iron. Chloritoid crystallized when these rocks experienced contact metamorphism with the Cretaceous granitic intrusions. The Middle to Upper Carboniferous strata are not present in this terrane, but Permian strata are much more widely developed than the older rocks. Clastic sediments predominate in the Permian, whilst cherts and volcanics are completely absent. Parts of the fine-grained clastic rocks show a distinct slaty cleavage and are worked for roofing-tiles.

Mesozoic sediments overlie the Upper Permian above a slight unconformity, corresponding to the lower part of the Lower Triassic. The Triassic strata consist of clastics deposited in a shelf to shallow sea environment and contain some tuff, but chert is absent. A few cycles of sedimentation are observed, and some beds contain abundant fossil bivalves and ammonites. The Jurassic is also mostly composed of clastic rocks without pelagic sediments, and similarly represents a few cycles of sedimentation. In the middle part of the Middle Jurassic limestone occurs yielding abundant corals and stromatoporoids. These lithologic features of Triassic to Jurassic formations of the South Kitakami terrane are in strong contrast to those of the same age in the other Japanese terranes, which are mostly pelagic, consisting of chert and fully marine, fine-grained sediments with few macro-fossils.

Plentiful volcanic and pyroclastic rocks of various compositions are characteristic of the Cretaceous formations of this terrane although the lowermost part consists of non-marine clastic rocks.

2.14   North Kitakami Terrane

The pre-Neogene geology of the North Kitakami terrane to the north of the Hayachine-Goyozan belt is not yet well understood since few studies have been made (SUGIMOTO, 1974). Strata, whose ages range from Permian to Jurassic, consist of slates, sandstones and cherts, and also occasionally embedded are fairly abundant masses and blocks of limestone and greenstone.

The lithofacial characteristics resemble those of the Permian to Jurassic formations of the subduction complexes of the Tamba, Mino and Chichibu terranes, Southwest Japan. Strata of similar age and lithology are also sporadically exposed in the southwestern peninsular part of Hokkaido to the north, under a cover of younger sediments and volcanics.

In the Pacific coast area, the Lower to Upper Cretaceous formations unconformably overlie the older rocks. The Cretaceous strata are rich in sandstones which are the products of fresh-water, shore and shallow sea environments. They contain abundant fossils such as hexacorals, gastropods and foraminifers.

2.15   Hokkaido

The island of Hokkaido is divided geologically into two parts: The southwestern peninsular and the central to eastern area. The Ishikari Lowland underlain by Quaternary sediments separates the above two. The pre-Neogene geology of the peninsular area is essentially the same as that of the North Kitakami terrane described above. On the other hand, in the central to eastern Hokkaido area the following four zones of pre-Neogene rocks are classified. These are, from west to east: the Ishikari-Kamuikotan, the Hidaka, the Tokoro-Toyokoro and the Nemuro belts. These all extend roughly in a N-S direction (Fig. 2.20).

2.15.1   Kamuikotan-Ishikari belt
On the western flanks of the axial range of Hokkaido, a tectonic zone, called the Kamuikotan belt has been developed. It is composed of high-pressure metamorphic rocks, ophiolitic rocks and serpentinite. It extends for about 300 km from the southern to the northern tip of the island, its maximum width in the central part reaches about 20 km. The northern extension of this belt may be traced northwards into Sakhalin.

The geology of the Kamuikotan belt is highly complicated, but, roughly speaking, an anticlinal structure may be recognized in the Horokanai and Kamuikotan gorge areas (ISHIZUKA et al., 1983). Its axial part is occupied by Kamuikotan metamorphic rocks and both limbs of the anticline by the Horokanai ophiolitic rocks. The Kamuikotan metamorphic rocks are glaucophanitic and have been known not infrequently to carry such character-istic minerals as glaucophane, lawsonite, jadeite and metamorphic aragonite. The latter three minerals were first reported in Japan from this terrane. The Kamuikotan belt is the largest serpentinite belt in Japan, and abundant ultrabasic and related rocks are associated with these metamorphic rocks. According to recent studies, however, this glaucophanitic terrane may not be a coherent regional metamorphic belt, but may consist of a number of

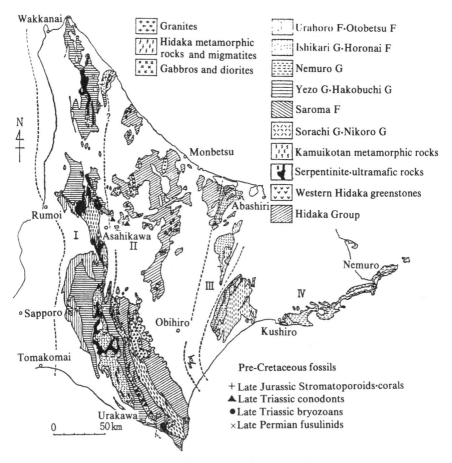

Fig. 2.20. Tectonic division of the older rock terranes in Central Hokkaido (modified after HIROKAWA *et al.*, 1978). I: Ishikari-Kamuikotan belt, II: Hidaka belt, III: Tokoro belt, IV: Nemuro belt, Not ornamented: Neogene to Quaternary.

tectonic blocks and masses of various sizes and rock types, which were pervasively invaded by serpentinite along their boundaries. Radiometric study of the metamorphic rocks gives a few clusters of age data. These features indicate that the terrane underlain by the Kamuikotan high-pressure rocks and serpentinite is a tectonic melange zone.

In the Horokanai district, central Hokkaido, the Kamuikotan meta-morphic rocks are overlain with a thrusted ophiolitic complex composed of pelagic sediments, hyaloclastics, pillowed basalts, foliated and massive amphibolites and peridotites. This complex called the Horokanai ophiolite has suffered low-pressure oceanic metamorphism.

The western and eastern sides of the Kamuikotan tectonic belt are covered by Cretaceous strata. The strata are classified into the Yezo Group (Aptian to Middle Campanian) and the Hakobuchi Group (late Campanian to Maastrichtian). Both groups consist predominantly of intercalated sandstones, siltstones, and mudstones. The lower part of the Yezo Group is represented by flysch-type alternations of sandstone and mudstone, while the middle to upper Yezo Group and the Hakobuchi Group consist of pebbly mudstone slump beds and exotic reef limestone. The limestone contains thick shells, hexacorals and algae. In the Yezo Group the sedimentary facies change from the western shallow-sea coarse-grained facies to the eastern off-shore fine-grained flysch-type facies (TANAKA, 1970; MATSUMOTO and OKADA, 1971). Furthermore, the Hakobuchi Group as a whole is of the regressive coarse-grained facies type (Fig. 2.21).

Plentiful fossil ammonites and bivalves are obtained from the Yezo and the Hakobuchi Groups. The detailed Upper Cretaceous biostratigraphy based on them gives the standard of the circum-Pacific Upper Cretaceous biostratigraphy (MATSUMOTO, 1959).

2.15.2   Hidaka metamorphic belt

The Hidaka metamorphic belt occupies the central ranges area of Hokkaido. It stretches for about 100 km from the Horoman district northwards to the Karikachi pass area in central Hokkaido. The northern

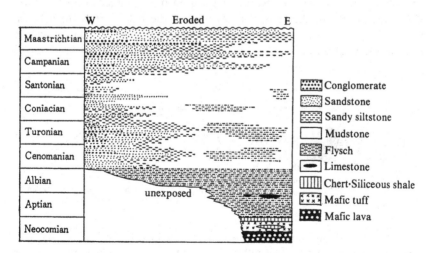

Fig. 2.21.   Lithological change of Cretaceous strata in Central Hokkaido (after MATSUMOTO and OKADA, 1971).

extension is obscured by the cover of younger sediments and volcanic rocks, but sporadic exposures of granitic rocks indicate that the belt may be traced northeastwards to the Okhotsk Sea coast area.

The Hidaka belt is subdivided into two parts: the western marginal zone and the main zone. The former is a belt of metamorphosed and tectonized ophiolitic complex and it consists of low-grade schists, amphibolites, meta-gabbros, ultrabasic metacumulates and peridotite tectonites.

A few representative E-W cross sections along the main zone show the following rock types appearing successively from west to east:

1) Granulites,
2) Amphibolites,
3) Intercalations of amphibolite and biotite gneiss,
4) Intercalations of biotite-hornblende gneiss and biotite gneiss,
5) Biotite-muscovite gneiss,
6) Mica schists,
7) Hornfels, and finally
8) Unmetamorphosed sedimentary rocks.

As shown in the sequence, the western half is dominated by basic rocks and the eastern half dominated by pelitic rocks. Metamorphic grade progressively increases from the eastern greenschist facies through the central amphibolite facies, to the westernmost granulite facies. The petrological features of the main zone may be interpreted as a section of crystalline basement derived from continental or island arc crust (KOMATSU et al., 1983) (Fig. 2.22).

Many plutonic masses, whose types of rock vary from olivine gabbro through diorite to granite were intruded into the metamorphic terrane. Migmatitic rocks in the central axial part of the belt were formed by the granite. Furthermore, peridotite masses have been emplaced particularly in association with ophiolitic rocks of the western zone. The Horoman mass in

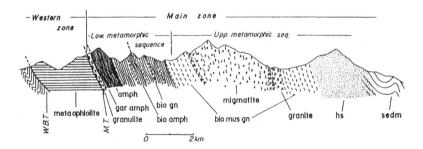

Fig. 2.22.   Cross section of the Hidaka metamorphic belt in the Koibokushu-shibichari-gawa to Satsunaigawa area. W.B.T.: western boundary thrust fault, M.T.: Hidaka main thrust falt. (after KOMATSU et al., 1983).

the southernmost part of the belt is the largest and consists of layered peridotite, ranging from dunite through lherzolite to plagioclase peridotite.

2.15.3 Tokoro-Toyokoro and Nemuro belts

To the east of the Hidaka belt the Tokoro-Toyokoro belt is developed. It is characterized by thick piles of greenstone beds along with subordinate tuffaceous sandstone, shale, limestone and chert. The strata are collectively called the Nikoro Group. The limestone yields stromatoporoids and hexa-corals of Late Jurassic age. The Nikoro Group is overlain by the Saroma Group consisting of sandstone, intercalated sandstone-shale and shale in ascending order. It appears to be of late Jurassic age (Oxfordian to Kimmeridgian), according to the bivalve *Ancella* obtained from the uppermost horizon.

The Upper Cretaceous to Paleogene Nemuro Group is exposed in the easternmost part of Hokkaido. In the lower part of the group pillowed alkali basalts are interbedded.

2.16 Paleogene Coal Bearing Strata

The Paleogene strata of Japan are sub-divided into two groups. The strata of the first group generally show a few cycles of sedimentation ranging from neritic-littoral to fluviatile deposits, the latter bearing coal seams. These strata unconformably overlie Cretaceous and older rocks, as well as the Cretaceous felsic igneous rocks, in various areas from the western to northern Kyushu and the western Chugoku regions, Southwest Japan, to the Joban and Kuji areas, Northeast Japan, and in the Ishikari and Kushiro districts, Hokkaido. The second group of Paleogene strata occurs in the Shimanto belt, which consists of flysch and mudstone facies sediments, the latter not infrequently containing blocks of greenstone.

The coal seams in the first group of Paleogene strata have been worked for a long time, and abundant information about their stratigraphy, litho-facies, geologic structure and fossils has been accumulated, also highly detailed geologic maps of the respective areas have been published.

2.16.1 Western Chugoku and north to western Kyushu

Coal-bearing Paleogene strata are developed in the coal fields of the Ube, Kokura, Chikuho, Karatsu-Sasebo, Sakito-Takashima, Miike and Amakusa-Koshikijima districts. Generally the lower beds are exposed in southeastern areas such as Amakusa, Takashima and Miike, while the upper beds occur successively in the northwestern districts, and finally the strata in the Karatsu-Sasebo district are of Miocene age.

Three cycles of sedimentation were developed, each commencing with non-marine purple to red basal conglomerate and sandstone, grading into the intercalations of marine and non-marine deposits and ending finally with non-marine muddy strata. The uppermost strata are covered by Miocene marine sediments, and the latter are further overlain in their turn by regressive coal-bearing deposits (Fig. 2.23). Furthermore, the contemporaneous heterotopic relations are observed: the strata in the southwestern areas are marine, while those of the same horizon in the northeastern areas are non-marine. In the non-marine formations a number of cyclothems are found, each ranging from sandstone-conglomeratic sandstone through siltstone

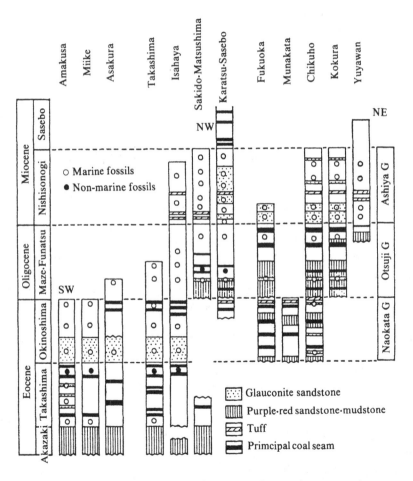

Fig. 2.23.   Lithostratigraphic changes of the Paleogene to Lower miocene strata in western to northern Kyushu and western Chugoku. G: Group. (after MATSUSHITA, 1949; MIZUNO, 1962 and 1963).

to coal-bearing mudstone (Fig. 2.24) (IWAHASHI and OHARA, 1959). Coal seams are generally less than 2 m in thickness, but a few in the Takashima coal field reach 4 m thick. Cyclic sedimentation has also been developed in the Paleogene strata of the Koshikijima and Tsushima districts.

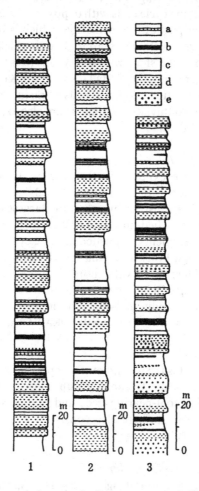

Fig. 2.24. Coal-bearing cyclothems in the Paleogene strata. 1: The Onga Formation of the Otsuji Group in the Kokura coal-field, Kyushu (after IWAHASHI and OBARA, 1959). 2: The Ikushumbetsu Formation of the Ishikari Group in the Yubari coal-field, Hokkaido (after SHIMOKAWA, 1963). 3: The Shakubetsu Formation of the Urahoro Group in the Kushiro coal-field, Hokkaido (after MABUCHI, 1962). a: Carbonaceous shale, b: Coal, c: Mudstone, d: Sandstone, e: Conglomerate.

2.16.2 Tohoku Province

In Northeast Japan the Paleogene strata have been substantially developed in the Joban and Kuji areas (Fig. 2.18). A sedimentary cycle is recognized in the Shiramizu Group of the former area. The cycle begins with littoral to fluviatile coal-bearing deposits, these grade into neritic sediments rich in fossil shells, and the cycle ends with deposition of massive mudstones. In particular, the lower part of the formation, which is of Oligocene age, is comprised of some dozen cyclothems, 10 to 35 m in thickness, consisting of conglomerate-sandstone and shale with coal seams (less than 2.5 m) or coaly shale beds. On the other hand, the Noda Group of the Kuji area is composed of four cycles of sedimentation, with coal beds being intercalated in the lower three cycles. Lacustrine to fluviatile sediments are also exposed in a few areas along the Kuzumaki Tectonic Zone.

2.16.3 Ishikari district, Hokkaido

Paleogene strata are exposed in the Rumoi, Ishikari, Yamabe and Hidaka areas, from north to south. Coal beds are widely developed particularly in the Ishikari district, which is the largest coal field in Japan. The Ishikari Group is classified into nine or eleven formations and consists mostly of non-marine sediments, deposited in three cycles, the facies ranging from fresh-water through brackish to marine. There are six principal coal-bearing formations intercalated with more than 200 coal beds. The coal-bearing beds are made up of a number of cyclothems, 10 to 35 m in thickness, composed of sandstone, sandy mudstone and mudstone with coal or coaly shale beds (Fig. 2.24) (TASHIRO, 1952; SHIMOKAWARA, 1963).

2.16.4 Kushiro district

The Kushiro district is the second of the important Paleogene coal fields in Hokkaido. The Oligocene Urahoro group consists of lower and upper fresh-water to brackish coal-bearing beds and middle shell-bearing neritic beds. The coal-bearing strata here also consist of a number of cyclothems (Fig. 2.24) (MABUCHI, 1962).

2.16.5 Outer Zone of Southwest Japan

Small areas underlain by Paleogene strata are sporadically distributed in the Sambagawa, Chichibu, Shimanto and Izumi terranes of the Outer Zone of Southwest Japan. The Kuma Group in western Shikoku has been relatively well studied (NAGAI, 1972). Its lower part is marine but the upper part is made up of two cycles of non-marine sediments. Other Paleogene strata include the Nakaoku Formation, similarly in Shikoku, the Mitake Formation, Kyushu, and the Wada Formation in the Tenryu district, central Japan.

REFERENCES

ASAMI, M. and M. HOSHINO, Staurolite-bearing schists from the Hongu-San area in the Ryoke metamorphic belt, central Japan, *J. Geol. Soc. Japan*, **86**, 581–591, 1980

BANNO, S., Petrologic studies on Sanbagawa crystalline schists in the Bessi-Ino district, central Shikoku, Japan, *J. Fac. Sci., Univ. Tokyo, Sec. II*, **15**, 203–319, 1964.

BANNO, S., K. YOKOYAMA, O. IWATA, and S. TERASHIMA, Genesis of epidote emphibolite masses in the Sanbagawa metamorphic belt of central Shikoku, *J. Geo. Scoc. Japan*, **82**, 199–210, 1976.

BANNO, S., C. SAKAI, and T. HIGASHINO, Pressure-temperature trajectory of the Sanbagawa metamorphis from garnet zoning, *Lithos*, **19**, 51–63, 1986.

ERNST, W. G., Y. SEKI, H. ONUKI, and M. C. GILBERT, Compaprative study of low-grade metamorphism in the California Coast Range and the outer metamorphic belt of Japan, *Geol. Soc. Am. Mem.*, **142**, 1970.

HARA, I., K. HIDE, K. TAKEDA, E. TSUKUDA, M. TOKUDA, and T. SHIOTA, Tectonic movement in the Sambagawa belt, *The Sambagawa Belt* (K. Hide, ed.), 307–390, 1977.

HASHIMOTO, M., Mineral facies of the Sengun metamorphic rocks of the Chugoku provinces, *Bull. Natn. Sci. Mus.*, **15**, 767–775, 1972.

HIDE, K., T. SUZUKI, and N. KASHIMA, The Sambagawa, Mikabu and Chichibu belts, *Guidebook for Excursion*, No. 5, Geol. Soc. Japan, 1977.

HIROI, Y., Subdivision of the Hida metamorphic complex, central Japan, and its bearing on the geology of the Far East in pre-Sea of Japan time, *Tectonophysics*, **76**, 317–333, 1981.

HIROKAWA, O. (Chief ed.), Geological Map of Japan, 1:1,000,000 (2nd Ed.), Geol. Surv. Japan, 1978.

IHIWATARI, A., A preliminary report on the Yakuno ophiolite in the Maizuru zone, inner southwest Japan, *"Earth Sciences"* (Chikyu Kagaku), **32**, 301–310, 1978.

ISHIWATARI, A., Granulite-facies metacumulates of the ;Yakuno ophiolite, Japan: Evidence for unusually thick oceanic crust, *J. Petrol.*, **26**, 1–30, 1985.

ISHIZUKA, H., M. IMAIZUMI, N. GOUCHI, and S. BANNO, The Kamuikotan zone in Hokkaido: Tectonic mixing of high-pressure and low-pressure metamorphic rocks, *J. Metamorphic Geol.*, **1**, 263–275, 1983.

IWAHASHI, T. and J. OHARA, A study on the stratigraphy and geological structure of the Kokura coal-field, northern Kyushu, Japan, *J. Geol. Soc. Japan*, **65**, 528–536, 1959.

KANMERA, K., Litho- and bio-facies of Permo-Triassic geosynclinal limestone of the Sambosan belt in southern Kyushu, *Pleont. Soc. Japan Special Paper*, **14**, 13–39, 1969

KANMERA, K. and H. NISHI, Accreted oceanic reef complex in Southwest Japan, *Accretion Tectonics in the Circum-Pacific Regions* (M. Hashimoto and S. Uyeda, eds.), 195–206, 1983.

KANO, H., Y. KURODA, K. URUNO, T. NUREKI, S. KANISAWA, T. MARUYAMA, H. UMEMURA, H. MITSUKAWA, N. SETO, Y. OHIRA, S. SATO, and N. ISSHIKI, Geology of the Takaniki district, Quadrangle Series 1:50,000, Geol. Surv. Japan, 1973.

KATTO, J., K. SUYARI, N. KASHIMA, I. HASHIMOTO, S. HADA, S. MITSUI, and I. AKOJIMA, Surface Geologic Map of Shikoku 1:200,000. Kochi Regional Forestry Office, 1977.

KOJIMA, G., K. HIDE and G. YOSHINO, The stratigraphical position of Kieslager in the Sanbagawa crystalline schist zone in Shikoku, *J. Geol. Soc. Japan*, **62**, 30–45, 1956.

KOMATSU, M., S. MIYASHITA, J. MAEDA, Y. OSANAI, and T. TOYOSHIMA, Disclosing of a deepest section of continental-type crust up-thrust as the final event of collision of arcs in Hokkaido, north Japan, *Accretion Tectonics in the Circum-Pacific Regions* (M. Hashimoto and S. Uyeda, eds.), 149–165, 1983.

MABUCHI, S., A study on sedimentation and tectogenic history of the Paleogene System of the Kushiro coal field, *Contrib. Inst. Geol. Paleont., Tohoku Univ.*, **56**, 1–42, 1962.

MARUYAMA, S. and M. YAMASAKI, Paleozoic submarine volcanoes in the high-P/T metamorphosed Chichibu System of eastern Shikoku, Japan, *J. Volcanol. and Geothermal Res.*, **4**, 199–216, 1978.

MARUYAMA, S., S. BANNO, T. MATSUDA, and T. NAKAJIMA, Kurosegawa zone and its bearing on the development of the Japanese islands, *Tectonophysics*, **110**, 47–60, 1984.

MATSUMOTO, T., *The Cretaceous System in the Japanese Islands*, Japan Soc. Prom. Sci., Tokyo, 1954.

MATSUMOTO, T., Zonation of the Upper Cretaceous in Japan, *Mem. Fac. Sci., Kyushu Univ., Ser. D.*, **9**, 55–93, 1959.

MATSUMOTO, T. and H. OKADA, Clastic sediments of the Cretaceous Yezo geosyncline, *Mem. Geol. Soc. Japan*, **6**, 61–74, 1971.

MATSUSHITA, H., Stratigraphical study of the Paleogene formations of north Kyushu, *Sci. Rept., Dept. Geol., Kyushu Univ.*, **3**, 1–57, 1949.

MIYASHIRO, A., Regional metamorphism of the Gosaisyo-Takanuki district in the central Abukuma plateau, *J. Fac. Sci., Univ. Tokyo, Sec. II*, **11**, 219–272, 1958.

MIYASHIRO, A., Evolution of metamorphic belts, *J. Petrology*, **2**, 277–311, 1961.

MIZUNO, A., Paleogene and Lower Neogene biochronology of west Japan, *J. Geol. Soc. Japan*, **68**, 640–648, 687–693, **69**, 38–50, 1962–1963.

MIZUTANI, S. and I. HATTORI, Hida and Mino: Tectonostratigraphic terranes in central japan, *Accretion Tectonics in the Circum-Pacific Regions* (M. Hashimoto and S. Uyeda, eds.), 169–178, 1983.

MURAKAMI, N., Cretaceous System, *Geology of Yamaguchi Prefecture*, 123–140, 1975.

NAKAZAWA, K., The Triassic System in the Maizuru zone, southwest Japan, *Mem. Fac. Sci., Kyoto Univ., Geol. Mineral.*, **24**, 265–313, 1958.

NISHIMURA, Y., E. NAKAMURA, and I. HARA, K-Ar ages of Sangun metamorphic rocks in Yamaguchi Prefecture and their geologic significance, *J. Japan. Assoc. Mineral. Petrol. & Econ. Geol.*, **78**, 11–20, 1983.

NOZAWA, T., Radiometro Age MAP OF japan, 1) Granitic rocks and 2) Metamorphic rocks, Geol. Surv. Japan, 1:2,000,000 Map Series, 16-1 & 16-2, 1977.

ONO, A., Zoning of the metamorphic rocks in the Takato-Shioziri area, Nagano Prefecture, *J. Geol. Soc. Japan*, **75**, 521–536, 1969.

ONO, A., Petrologic study of the Ryoke metamorphic rocks in the Takato-Shiojiri area, central Japan, *J. Japan. Assoc. Mineral. Petrol. & Econ. Geol.*, **72**, 453–468, 1977.

ONUKI, Y., Geology of the Kitakami mountainland, Iwate Prefectural Agency, 1956.

SAITO, Y. and M. HASHIMOTO, South Kitakami region: An allochthonous terrane in Japan, *J. Geophys. Res.*, **87**, 3691–3696, 1982.

SAKAGUCHI, S., Geology of the Tamba district, Memorial Publ. in Comm. of Prof. Sakaguchi's Retirem., 1974.

SEKI, Y., Glaucophanitic regional metamorphism in the Kanto mountains, central Japan, Japan. *J. Geol. Geogr.*, **29**, 233–258, 1958.

SHIDO, F., Pluatonic and metamorphic rocks f the Nakoso and Iritono districts in the central Abukuma plateau, *J. Fac. Sci., Univ. Tokyo, Sec. II*, **11**, 131–217, 1958.

SHIMOKAWARA, T., Geology and structural development of the Yubari coalfield, Hokkaido, Japan, The Hokkaido Association of Coal Mining Technologists, 1963.

SUGIMOTO, M., Stratigraphical study in the outer belt of the Kitakami masslf, Northeast Japan, *Contrib. Inst. Geol. Paleont., Tohoku Univ.*, No. 74, 1974.

SUYARI, K., On the lithofacies and the correlation of the Izumi Group of the Asan mountain range, Shikoku. *Sci. Rep., Tokushima Univ., Ser. II*, Spec. Vol, 6, 489–495, 1973.

SUZUKI, M., An occurrence of "eclogitic rock" in the Hida metamorphic belt, *J. Japan. Assoc. Mineral. Petrol. & Econ. Geol.*, **68**, 372-382, 1973.

TAGIRI, M., Metamorphism of Paleozoic rocks in the Hitachi district, southern Abukuma plateau, Japan, *Sci. Rept. Tohoku Univ., Ser. III*, **12**, 1-67, 1973.

TAMBA BELT RESEARCH GROUP, Geosynclinal facies of the Tamba belt, Southwest Japan, Assoc. for Geol. Collab. Japan Monogr., **19**, 13-23, 1975.

TANAKA, K., Sedimentation of the Cretaceous flysch sequence in the Ikushunbetsu area, Hokkaido, Japan, *Rept. Geol. Surv. Japan*, **236**, 1-107, 1970.

TASHIRO, S., Deposition of coal and cycle of sedimentation, *J. Geol. Soc. Japan*, **58**, 529-536, 1952.

TOKUYAMA, A., Die obertriadische Molasse in Mine-Gebiet Westjapans, *J. Geol. Soc. Japan*, **64**, 454-463 & 537-550, 1958.

TOYOHARA, F., Geologic structure from Sangun-Yamaguchi zone to "Ryoke zone" in eastern Yamaguchi Prefecture, *J. Geol. Soc. Japan*, **82**, 99-111, 1976.

YANAGIMOTO, Y., Stratigraphy and geological structure of the Paleozonic and Mesozoic formations in the vicinity of Kuzuu, Tochigi Prefecture, *J. Geol. Soc. Japan*, **79**, 441-451, 1973.

CHAPTER 3

Chapter 3

ULTRAMAFIC ROCKS AND GABBROS

Numerous lenticular ultramafic masses are distributed in the metamorphic and tectonic belts of the Japanese Islands. These rocks are commonly highly serpentinized, but the primary minerals are occasionally preserved.

Systematic studies of Japanese ultramafic rocks are not yet completed, although those occurring in the Sangun, Sambagawa and Hidaka belts are relatively well studied.

3.1 The Sangun Belt

Figure 3.1 shows the distribution of ultramafic masses in the Sangun and Maizuru belts. According to their internal structures, they are classified into massive and layered types (ARAI, 1975). The massive types are represented by the Tari-Misaka mass in Tottori Prefecture and the Oeyama mass of Kyoto Prefecture. The layered types occur in the Ochiai-Hokubo

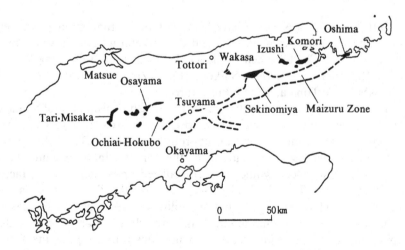

Fig. 3.1.   Ultramafic masses in the Sangun and Maizuru terranes.

mass in Okayama Prefecture and the Oshima mass of Fukui Prefecture. Generally, a zonal arrangement is observed, the massive bodies occurring in the northern zone and the layered masses in the southern zone.

The rocks were altered not only by serpentinization, but also by later contact metamorphism due to Cretaceous granitic intrusions. Analysing the repeated metamorphic effects shows that the original peridotites would have been dunite, harzburgite and chromitite and relatively poor in clinopyroxene. The innermost part of the contact aureole of the Tari-Misaka mass is characterized by the association orthopyroxene-olivine. The orthopyroxene is very poor in Ca compared with that of the original peridotite.

The layered-type masses contain lherzolite, wehrlite and clinopyroxenite in addition to the principal dunite, and clinopyroxene is also one of their main constituents. Tremolite appears in the contact metamorphosed rocks of these layered-type peridotites.

The massive-type peridotites do not show distinct layering, and do not vary in their mineral parageneses, additionally, their olivine and clinopyroxene are rich in Mg. These features suggest that the peridotites are residual parts of the partially melted mantle material. On the other hand, the layered type rocks exhibit cryptic layering of chemical composition of the minerals, and also show relatively wide compositional variations of chromite and spinel. It is considered, therefore, that the layered peridotites are the products of mixtures of partially melted lherzolite and solidified magma.

3.2 Sambagawa Belt and Contiguous Areas

Ultramafic masses are relatively abundant in the Sambagawa belt proper, the southern marginal zone and the Mikabu terrane, however they are scarce in the Chichibu belt to the south. Particularly well studied areas are the Higashiakaishi mass in Shikoku (Fig. 3.2) (BANNO et al., 1976) and the masses in the Toba area, Mie Prefecture.

Dunite is the principal rock type of the Higashiakaishi peridotite mass, although other rock types such as wehrlite, clinopyroxenite, garnet clino-pyroxenite and chromitite occur in subordinate amounts. Detailed analyses of mineral parageneses, particularly of the garnet clinopyroxenites have indicated that the rocks would have crystallized under amphibolite facies conditions (500–600°C, 7–13 kb). But, similarly detailed studies on the spinel-bearing clinopyroxenite and granulite in small peridotite masses embodied in the Irazu amphibolite mass have shown that the peridotite-amphibolite complex would have been previously under granulite facies conditions (700–900°C, 5–8 kb). Furthermore, the peridotite and amphibolite are considered to have been the products of fractional crystallization of olivine tholeiite magma under lower crust or upper mantle conditions

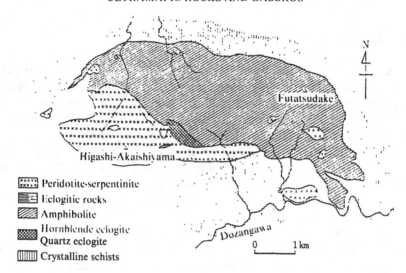

Legend:
- Peridotite-serpentinite
- Eclogitic rocks
- Amphibolite
- Hornblende eclogite
- Quartz eclogite
- Crystalline schists

Futatsudake

Higashi-Akaishiyama

Dozangawa

0   1 km

Fig. 3.2. The Higashi-Akaishi peridotite mass and the Irazu amphibolite mass (simplified after BANNO *et al.*, 1976).

(1300° C, 13 kb), the amphibolite having been originally gabbro.

Ultramafic rocks of the Mikabu terrane are classified into those rich and poor in hornblende (NAKAMURA, 1971; YOKOYAMA, 1976). The former are olivine hornblendite and hornblende gabbro, while the latter are dunite, wehrlite and plagioclase peridotite. The Mikabu rocks are relatively poor in Mg. $MgO/(Fe_2O_3+FeO+MgO)$ of the ultramafic rocks is less than 0.8, and the Fo content of the olivine is lower than 90%. The ultramafic rocks have been affected by the pumpellyite-actionolite facies metamorphism in a similar way to the country rocks and frequently carry such minerals as chrysotile, hydrogarnet, chlorite and secondary clinopyroxene with acicular habit. However, the parageneses of the inferred primary minerals indicates that the ultramafic rocks would have been formed as cumulates derived from picritic basalt magma under fairly high temperatures (1000–1200° C).

In the Shimanto belt ultramafic rocks are mostly emplaced into the Paleogene Setogawa Group, Shizuoka district, and the Mineoka Group of the Boso Peninsula. The primary rock types are inferred to be plagioclase-bearing harzburgite and dunite. A highly magnesium olivine (Fo_{94}) has been found in the dunite of the Mineoka area. Awaruite was discovered, for the first time in Japan, from serpentinites of the Mineoka and Chichibu terranes.

3.3   Hidaka and Kamuikotan Belts

Of numerous ultramafic bodies emplaced in the Hidaka metamorphic belt, the Uenzaru mass in the northern area, and the Horoman mass in the southern area, have been well studied. The Uenzaru mass consists of lherzolite and dunite along with minor pyroxenite and little orthopyroxene-bearing rocks (KOMATSU, 1975). The rocks contain primary minerals with relatively large crystals and secondary minerals have been shattered to fine grains. The primary pyroxene is rich in Al_2O_3 and has exsolved lamella of plagioclase and spinel in the recrystallized parts. The rocks of the primary assemblages are considered to have been formed under conditions of lower crust to upper mantle.

The Horoman mass shows a distinct layered structure, consisting of a few units of layers (NAGASAKI, 1966). A unit undergoes changes in its rock types, from lower to higher layers, as follows: spinel-bearing dunite, spinel lherzolite and plagioclase peridotite. Furthermore, the lower units are richer in olivine and poorer in plagioclase than the higher units. Therefore, the Horoman mass has a doubly layered structure.

The Kamuikotan belt is the largest serpentinite belt in Japan. In the Horokanai district ultramafic rocks constitute the basal part of an ophiolite complex, with gabbro and basalt lying successively upon them (p. 45). Similar ophiolitic complexes of peridotite-gabbro-basalt are also reported in the Maizuru Belt.

3.4   Other Areas

Large peridotite masses are exposed in the Hayachine and Miyamori districts of the Kitakami Mountains, Northeast Japan. They consist primarily of dunite, wehrlite, hornblendite and hornblende gabbro, but little ortho-pyroxene-bearing rocks. The Miyamori mass has been affected by contact metamorphism of the Cretaceous Tono Granodiorite and has been recrystalliz-ed into various hornfelses bearing Mg-rich minerals.

Cortlanditic masses are not infrequently found in the gneiss regions of the Ryoke and Abukuma belts, but it is uncertain whether the observed mineral composition is primary or not. Detailed studies have not been made on these mafic and ultramafic rocks.

3.5   Summary

Ultramafic rocks occurring in the regional metamorphic terranes of Japan are mostly recrystallized under the same conditions as those of the country rocks. It is suggested that they were emplaced before the final stage of metamorphism. However, the primary minerals and textures are widely preserved in these partly recrystallized peridotites. The analyses of parageneses and textures suggest that the original rocks are classified into the following three groups. The first group belongs to the lower part of the ophiolitic complexes such as those of the Kamuikotan and Maizuru belts. They may have been derived from mantle under oceanic crust. The second group are represented by rocks of the Mikabu and Shimanto belts. They are cumulates derived from basaltic magmas. The last group occur in the Hidaka and Sambagawa belts. Before emplacement, they would have been under the granulite facies conditions. Conditions, however, would have varied from mass to mass even in the same metamorphic terrane.

3.6   Rocks Associated with Ultramafic Rocks

Rodingite occurs widely, particularly in association with highly serpentinized rocks. These show mineral parageneses in harmony with that of the country serpentinites. Other rocks associated with serpentinites are albitite and jadeitite. Jadeitite has been reported from the Omi, Sekinomiya, Wakasa and Sonogi districts mostly in the Sangun belt.

Serpentine sandstone and serpentinite conglomerate are intercalated in the Neogene strata of, for example, Hokkaido, the Miura Peninsula and Okayama Prefecture (e.g. UCHIDA and ARAI, 1978).

3.7   Gabbros

A few gabbroic intrusives occur in the Cenozoic terranes of Japan. Of these, those of the Muroto district, Kochi, and of the Sanogawa area, Shizuoka are mentioned here.

The Muroto gabbro is a stock intruded into the Eocene Muroto Group. The principal rock is olivine gabbro with a little orthopyroxene, but neither pigeonite nor quartz occurs. The Sanogawa gabbro consists of six gabbroic to dioritic stocks emplaced in the Miocene Fujigawa Group. Olivine-bearing, two pyroxene gabbro, diorite and minor granophyre are the main rock types. They contain orthopyroxene, hornblende and quartz and exhibit a calc-alkalic trend of differentiation. A similar gabbroic mass called the Koyama gabbro complex, has been intruded into the Lower Miocene Susa Group in northwestern Yamaguchi Prefecture. Various rocks bearing orthopyroxene, hornblende and quartz are observed in the complex.

REFEREMCES

ARAI, S., Petrology of alpine-type ultramafic complexes in the Sengun zone, western Japan, D. Sci. Thesis, Univ. Tokyo, 1975.

BANNO, S., YOKOYAMA, K., IWATA, O, and TERASHIMA, S., Genesis of epidote amphibolite masses in the Senbagawa metamorphic belt of central Shikoku, *J. Geol. Soc. Japan*, **82**, 199–210, 1976.

KOMATSU, M., Recrystallization of high-alumina pyroxene peridotite of the Uenzaru area in Hidaka province, Hokkaido, Japan, *J. Geol. Soc. Japan*, **81**, 11–28, 1975.

NAGASAKI, H., A layered ultrabasic complex at Horoman, Hokkaido, Japan, *J. Fac. Sci., Univ. Tokyo, II*, **16**, 313–346, 1966.

NAKAMURA, Y., Petrology of the Toba ultrabasic complex, Mie Prefecture, central Japan, *J. Fac. Sci., Univ. Tokyo, II*, **18**, 1–51, 1971.

UCHIDA, T. and ARAI, S., Petrology of ultramafic rocks from the Boso peninsula and the Miura peninsula, *J. Geol. Soc. Japan*, **64**, 561–570, 1978.

YAJIMA, T., The Sanogawa gabbro-diorite complex, *Sci. Rept., Saitama Univ., B*, **5**, 199–230, 1970.

YAJIMA, T., Petrology of the Murotomisaki gabbroic complex, *J. Japan. Ass. Min. Pet. Econ. Geol.*, **67**, 218–241, 1972.

YAMAZAKI, T., Petrology of the Koyama calc-alkaline intrusive comples, Yamaguchi Prefecture, Japan, *Sci. rept., Tohoku Univ., III*, **10**, 99–150, 1967.

YOKOYAMA, K., Ultramafic and related rocks in the Sanbagawa metamorphic belt, D/Sci. Thesis, Univ. Tokyo, 1976.

CHAPTER 4

GRANITES AND RHYOLITES

Over 70% of the Japanese granitic and associated rhyolitic rocks are of Cretaceous to Paleogene age, while the remainder are either of pre-Jurassic or Neogene age. The granitic rocks are grouped into two distinctive types, on the basis of amount and type of opaque oxide minerals included in them (Fig. 4.1).

The magnetite-series granitic rocks contain 0.2 to 2.0% by volume of magnetite, in addition to small amounts of ilmenite, hemetite, pyrite (and/or chalcopyrite), sphene and epidote group minerals. The biotites and/or hornblendes are commonly high in the Fe^{3+}/Fe^{2+} and Mg/Fe ratios and low in the indices of refraction. On the other hand, rocks of the ilmenite-series are practically free from oxide opaque minerals and contain less than 0.2% by volume of ilmenite. The ratios Fe^{3+}/Fe^{2+} and Mg/Fe of the biotites are low and the refractive indices are high (ISHIHARA, 1977).

The total iron content is not significantly different in the granitoids of both series, but iron oxide minerals/Fe–Mg silicates is higher in the magnetite-series rocks. This suggests that the oxygen fugacity would have been higher in the magnetite-series rocks than in the rocks of the ilmenite-series (Fig. 4.2). Furthermore, the magnetite series have larger amounts of basic rocks and are poorer in lithophile elements such as K, Rb, F, Li, Sn and others than the ilmenite series. $\delta^{18}O$ and $^{87}Sr/^{86}Sr$ initial ratios are lower and $\delta^{34}S$ is higher in the former than the latter series rocks (Fig. 4.3). Thus the ilmenite-series granitic rocks may have been formed under the influence of crustal materials.

4.1 Pre-Jurassic Granitoids

Granitic rocks of pre-Jurassic age are limited in their distribution to large fault zones such as the Kurosegawa, the Maizuru and the Nagato belts, all in Southwest Japan. Only the Funatsu granitoids of the Hida terrane have a relatively wide areal extent. Quartz diorites to granodiorites are predomi-

Fig. 4.1. Distribution of the magnetite-series and ilmenite-series granitoids (after ISHIHARA, 1977). The Funatsu granite and the granitoids of the Green Tuff region are not shown.

nant and commonly associated with gabbros and ultramafic rocks. Thus the Funatsu granitoids as a whole are rich in Ca and low in K/Na. They have been intruded into sedimentary rocks of Carboniferous to Permian age and were subsequently overlain by the Jurassic Tetori Group. A part of the intrusives are harmonious and gradational to the country rocks, but the larger part is not concordant with the host. The radiometric ages are mostly about 180 Ma, and the time of the final consolidation is considered to have been Early Jurassic. Two petrographical types have been classified: the Shimonomoto and the Funatsu types. The Shimonomoto type is relatively basic in composition, ranging from granodiorite to diorite. On the other hand, biotite monzogranite is the most predominant rock of the Funatsu

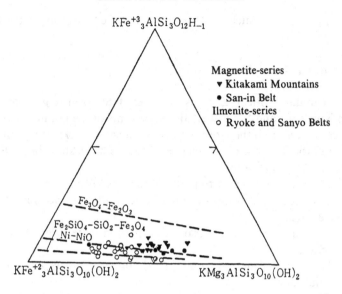

Fig. 4.2. Composition of biotites of the magnetite-series and ilmenite-series granitoids (ISHIHARA, 1977).

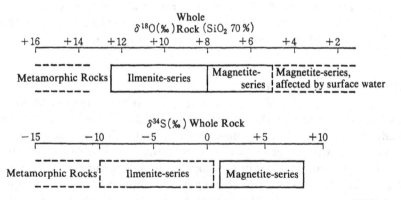

Fig. 4.3. Ranges of $\delta^{18}O$ (‰) (SiO 70%) and $\delta^{34}S$ (‰) of whole rocks of the magnetite-series and ilmenite-series granitoids. Rocks of Southwest Japan alone are shown in the diagram (after SASAKI and ISHIHARA, 1979).

type. Not infrequently augen gneiss has been developed along the peripheries of the Funatsu type intrusive masses.

The Funatsu Granites belong mostly to the magnetite-series and the following geochemical characteristics have been observed: rich in Na_2O as compared with the average Japanese granitic rock, rich in MgO, but poor in the lithophile elements such as Rb and Sn, and the Sr isotope initial ratio is about 0.7050.

4.2   Cretaceous Felsic Igneous Rocks

The Tanakura Shear Zone is considered to form a geologic divide between Northeast and Southwest Japan in terms of the granitic petrology.

As compared with the Cretaceous felsic igneous rocks of regions to the southwest of the fault zone, the igneous rocks of northeastern Japan have the following contrasting features:

1)   The ratio of volcanic to plutonic rocks is low,

2)   Chemical composition is more basic and the proportion of gabbro to granite is high,

3)   The magnetite-series granitoids predominate over those of the ilmenite-series,

4)   Rocks are generally poor in lithophile elements and low in Sr isotope initial ratio,

5)   The amount of magnetite in granitoids vaguely increases from the Sea of Japan areas, to the Pacific regions of Northeast Japan, while in Southwest Japan it does so distinctly towards the Sea of Japan areas, and

6)   The Mo–Pb–Zn metallogenic provinces, relative to the W–Cu provinces, are characteristic of the Pacific areas of Northeast Japan, whereas in Southwest Japan they are developed on the Sea of Japan side.

4.2.1   Northeast Japan

a)   *Kitakami Mountains*

A number of granitic masses are exposed in the Kitakami Mountains district. The largest of these is the Tono mass, 615 km^2 in areal extent. Radiometric ages of the granitoids are about 120 Ma and suggest, in conjunction with the geological evidences, that the time of their intrusion was Early Cretaceous. One of the most remarkable features of the Kitakami granitoids is that alkalic rocks having more potassium feldspar than quartz have been closely associated with ordinary calc-alkalic rocks which are richer in quartz than in potassium feldspar.

The granitoid masses are classified into six groups respectively exposed in each tectonic zone roughly trending in a NNW-SSE direction, as shown in Fig. 4.4 (KATADA *et al.*, 1974). Comparatively large intrusives occur particularly in Zone I, Zone III and Zone V. They consist mainly of rocks ranging from granodiorite to tonalite. On the other hand, granitic masses of Zone IV and Zone VI are accompanied by gabbro or diorite and contain, in part, alkalic rocks rich in potassium feldspar. Volcanic and pyroclastic rocks associated with and intruded into by the granitoids are mostly distributed in

Zone I and Zone IV. They are dacitic to andesitic in composition.

The granitoids of the Kitakami Mountains are generally of the magnetite-series, though rocks of the ilmenite-series are scarce (Fig. 4.4). The Sr isotope initial ratio is fairly low, around 0.7045. $\delta^{18}O(SiO_2\ 70\%)$ is larger than 8‰ in the largest Tono mass, and $\delta^{34}S$ of the whole rock has been measured as 2‰.

b) *Abukuma Plateau*

Granitic rocks are widely exposed in the Abukuma Plateau. They are closely associated with regional metamorphic rocks, particularly in the southern part of the plateau, where the rocks not infrequently show schistose structures. The principal rock type is hornblende-biotite granodiorite, although small amounts of biotite monzogranite and two-mica monzogranite are observed in some areas. The radiometric ages are about 90 Ma, but much older age data have been obtained for some rocks. The chronology of the Abukuma granitoids is not yet fully understood.

Most of the Abukuma granitoids are of the ilmenite-series, but in the eastern part of the area, small amounts of the magnetite-series rocks are found. Roughly speaking, therefore, the magnetite-series rocks increase in volume eastwards. The complicated distribution pattern of the two rock series is observed in the southern part of the plateau where regional metamorphic rocks are widely developed. The proportion of the ilmenite- to magnetite-series rocks does not vary systematically from mass to mass.

The Abukuma granitoids are poor in Rb, Sn and F, and their Sr isotope initial ratios are about 0.705. $\delta^{18}O(SiO_2\ 70\%)$ of the magnetite-series rocks is larger than 10‰ and $\delta^{34}S$ is +4‰.

4.2.2 Southwest Japan

In Southwest Japan the terranes of Cretaceous to Paleogene felsic igneous rocks are divided into three belts running parallel to each other in a ENE-WSW direction. They are, from south to north, the Ryoke belt, the Sanyo belt and the San-in belt. The Ryoke belt is associated with the Ryoke low P/T regional metamorphic rocks.

In the Sanyo and San-in belts, massive monzogranite and granodiorite are the main rock types, while schistose granodiorite is one of the principal facies in the Ryoke belt. The magnetite-series granitoids and their magnetite contents increase northwards from the Ryoke to the San-in belts. The following oxygen and sulphur isotope data have been obtained for three groups.

	Ryoke	Sanyo	San-in
$\delta^{18}O(SiO_2\ 70\%)$	+11‰	+9‰	+7‰
$\delta^{34}S$	−4‰	−3‰	+5‰

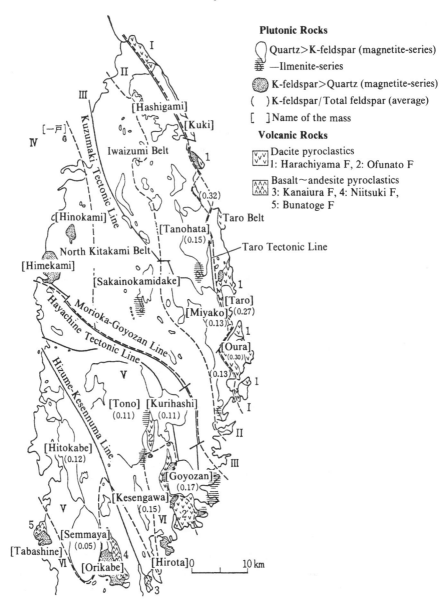

Plutonic Rocks

◯ Quartz>K-feldspar (magnetite-series)

—Ilmenite-series

◉ K-feldspar>Quartz (magnetite-series)

( ) K-feldspar/Total feldspar (average)

[ ] Name of the mass

Volcanic Rocks

Dacite pyroclastics
1: Harachiyama F, 2: Ofunato F

Basalt~andesite pyroclastics
3: Kanaiura F, 4: Niitsuki F,
5: Bunatoge F

Fig. 4.4. Zonal classification of Cretaceous granitoids in the Kitakami mountains. I to VI indicate the zones defined by KATADA *et al.*, 1974. F: Formation.

The Sr isotope initial ratio also decreases northwards from higher than 0.710 in the Ryoke belt to lower than 0.706 in the San-in belt, as shown in Fig. 4.5.

The radiometric ages of the granitic rocks become systematically younger from northern Kyushu eastwards. In the respective districts Cretaceous strata are developed in association with each K–Ar age group of the granitoids. Therefore, the eastward migration in the upheaval of the granitic masses and the concomitant development of sedimentary basins is suggested (MATSUMOTO, 1977).

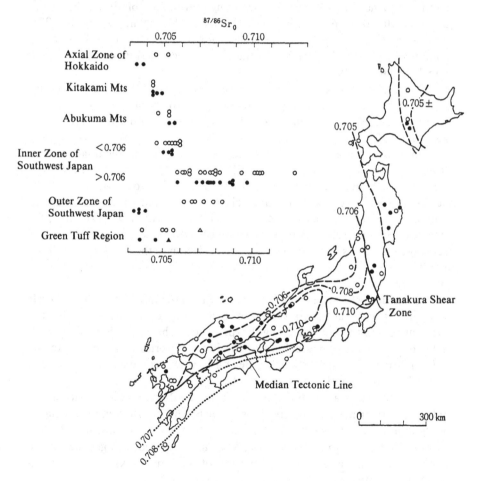

Fig. 4.5. Regional variation of the initial Sr ratio of plutonic rocks. Broken and dotted lines are the respective contours of Cretaceous to Paleogene and Neogene rocks. Open circle: granitic rocks, Solid circle: gabbroic and dioritic rocks, Open and solid triangles: rocks of the Green Tuff region in Southwest Japan.

a) *Ryoke belt*

Detailed field studies in central Japan show that the Ryoke granitic rocks are classified into two groups: those older and younger than the Nohi Rhyolites (p. 28) (YAMADA *et al.*, 1977). The most widely developed member of the older granitoids is the Tenryukyo granitoid. These rocks are composed mainly of hornblende-biotite granodiorite and monzogranite, and they include abundant basic xenoliths and are strongly schistose and porphyritic. The older granitoids are generally in harmonious relation with the surrounding Ryoke metamorphic rocks (biotite gneiss). On the other hand, the principal rock types of the younger group are biotite±hornblende granodiorite to monzogranite and muscovite±biotite monzogranite. The normative C and en/(en+fs) decrease northwards, but or+ab and the contents of Rb and Sr increase in the same direction (Fig. 4.6) (ISHIHARA and TERASHIMA, 1977).

The Ryoke granitic rocks in the other districts to the west are generally similar to those of central Japan. But some in northern Kyushu are of the magnetite-series and are comparatively low in the Sr isotope initial ratio, i.e. less than 0.706.

In central Japan, the schistose structures developed regionally in the Ryoke granitoids are roughly parallel to the direction of the Median Tectonic Line and become more distinct towards it. However, the strongly schistose rocks are not distributed close to the Median Line in western Japan. They are exposed in areas about 100 km distant from the line, for example, in the Yanai district. It is suggested that the formation of the schistose structures does not bear a genetic relation to the development of the Median Line.

b) *Sanyo belt*

Granitic rocks are exposed in wide areas to the north of the Ryoke belt and are called the Sanyo granitoid belt. They are represented by the Hiroshima granites in the Chugoku province. The rocks commonly lack schistose structure. They belong to the ilmenite-series and are rich in the lithophile elements such as Rb, Sn, Be and F, whose contents vary in relation to the SiO_2 content of the rocks (Fig. 4.7).

The Sanyo granitoids occur generally as disharmonious batholithic intrusives; their main rock type consists of biotite monzogranite associated with minor hornblende-biotite granodiorite and tonalo-diorite. Monzo-granites of the Sanyo belt are rich in K_2O, as compared with granites of the other two belts, and a small amount of syenitic rocks also occur in association with them.

c) *San-in belt*

Rock types of the San-in belt granitoids are similar to those of the Sanyo belt. The representatives are the Ogano or Tottori granites. In the

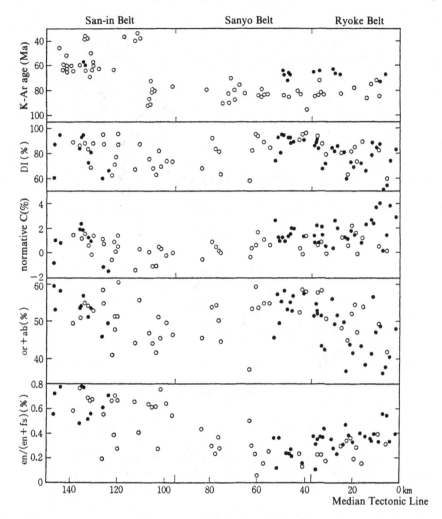

Fig. 4.6. N-S variation in various features of granitoids (modified after ISHIHARA and TERASHIMA, 1977). Open circle: central and eastern Chugoku, Solid circle: western Chubu.

peripheral parts of batholithic masses of the San-in granitoids fine-grained granophyres occur, which are occasionally accompanied by pegmatites and contain abundant xenoliths derived from the country rocks.

Radiometric ages have been determined as 65 to 60 Ma (Paleogene), being clearly younger than those of the Sanyo granitoids, and actually, in the northern Hiroshima district the former intrude into the latter. Furthermore, the magnetite-series rocks predominate in the San-in belt, in contrast to the

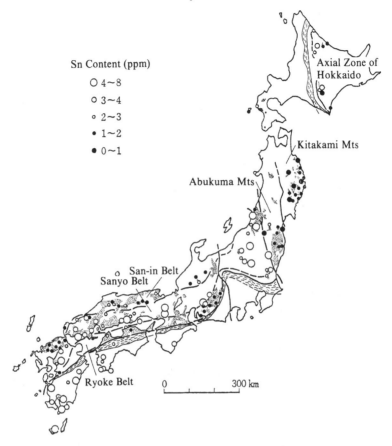

Fig. 4.7. Principal granitoid masses (stippled) and their Sn content (after ISHIHARA and TERASHIMA, 1977). Wavy ruled area: Sambagawa and Kamuikotan high P/T belts.

Sanyo belt. Thus the following correlations are recognized between geochemical and chronological features in the granitoids of both belts (Fig. 4.8) (ISHIARA, 1978).

	San-in	Sanyo
Age	Paleogene	late Cretaceous
Magnetite-series rocks/ilmenite-series rocks	higher	lower
Mo/W of the associated ore deposits	higher	lower

Fig. 4.8. Magnetite-series and ilmenite-series granitoids in the central part of Chugoku province. Solid circle magnetite-series rocks. The bigger the circle, the larger the amount of magnetite. Open circle: ilmenite-series rocks, Cross with Fe: principal placer iron deposit. Numerals indicate the K-Ar ages after SHIBATA and ISHIHARA, 1974, etc. Broken lines show the regional variation of $\delta^{18}O$ of quartz.

d) *Volcanic rocks of the Inner Zone of Southwest Japan*

Abundant volcanic and pyroclastic rocks, including felsic welded tuffs, are widely exposed in the Inner Zone of Southwest Japan. They are stratigraphically divided into five groups, whose ages range from late Cretaceous to Paleogene (Table 4.1) (TANAKA and NOZAWA, 1977). Rocks of stages I and II are developed mostly in the western areas and are andesite to dacite in composition. The Shimonoseki Subgroup of stage I is characterized by fine-grained red tuffs, which have been quarried as raw material for inkstone. Rocks of stage III are the most abundant and are also the most wide spread in exposed area, for example the Nohi Rhyolites in central Japan. The principal rock type is rhyolitic welded tuff with minor andesite. It is worthy of note that in stage V volcano-plutonic complexes have been formed in a few districts in connection with the development of caldera-like

Table 4.1   Volcanic rocks of the Inner Zone of Southwest Japan.

	Time of eruption	Western Chugoku	Central Chugoku	Eastern Chugoku ~Western Kinki	Eastern Kinki	Chubu	Kanto~Uetsu
Paleogene	V	Tamagawa (100 km³ central)	Kawauchi				
Paleogene	IV		Sakugi (800 km³ central)	Tenkadaiyama (50 km³ central) Kongodoji (10 km³ central)		Futomiyama (1250 km³ fissure, central) Oamamiyama Ishizaka	Katashina Honetateyama
Late Cretaceous	III	Abu (4500 km³ fissure)	Takada (15000 km³ fissure)	Yatagawa (1000 km³ ?) Ikuno (10000 km³ fissure) Arima (1600 km³ fissure) Aioi (3700 km³ fissure)	Koto (>1000 km³ fissure?)	Nohi (16000 km³ fissure) Omodani	Okunikko (>1000 km³ fissure) Tagawa
Late Cretaceous	II	Shunan (800 km³ fissure)	Kisa, upper (2000 km³ ?) Kisa, lower (1000 km³ central)	Hiromine (300 km³ fissure)			
Late Cretaceous	I	Shimonoseki (2400 km³ central)		Sennan (5000 km³ + fissure) Sasayama (100 km³+?)		Arimine	

Volume and type of eruption are shown in parentheses. Mostly after Tanaka and Nozawa (1977).

subsidence structures. The most well-known is the Tamagawa Cauldron, northeastern Yamaguchi Prefecture (MURAKAMI, 1973). It has an eliptical outline, 14×7 km^2, stretching in a NE-SW direction. The volcanic activity occurred in three stages: the first consisted of andesite lavas, the second of dacitic tuff breccia with andesite breccia, and the third of rhyo-dacitic tuff breccia with andesite dikes extending in the NE-SW direction. Plutonic, but fairly shallow, intrusions have taken place along the caldera wall and are considered to have occurred under the influence of meteoric water (MURAKAMI, 1969).

4.3   Neogene Granites

The Neogene granitic rocks are exposed in two distinctive belts, both running parallel to the direction of the Japanese island arc. The first forms the outer arc plutonic belt situated between the Japan trench and the volcanic front, and the second forms the inner arc plutonic belt located about 360 km from the trench axis (Fig. 4.9). Radiometric ages are mostly determined at about 15 Ma, but rocks of the former belt are slightly younger than rocks of the latter belt.

Neogene rocks of the two belts show respectively the following contrasting petrological and geochemical features (Fig. 4.10).

Outer arc plutonic belt
    1)   Associated with rare volcanics
    2)   Generally acidic in composition and of the ilmenite-series
    3)   Rich in lithophile elements, and high in K/Na and low in K/Rb
    4) $\delta^{18}O$ (SiO_2 70%) is larger than 9‰. $\delta^{34}S$ has negative value. High Sr isotope initial ratio

Inner arc plutonic belt
    1)   Generally basic (tonalitic to gabbroic) and mostly of the magnetite-series
    2)   Considerably high proportion of volcanic/plutonic rocks
    3)   Poor in lithophile elements
    4) $\delta^{18}O$ less than 7‰, $\delta^{34}S$ larger than 0. Low Sr isotope initial ratio

a)   *Axial zone of Hokkaido*

Some small granitic masses of the ilmenite-series are exposed in areas to the east of the Hidaka metamorphic belt; these mostly consist of biotite granite and granodiorite. Compared to granitic rocks of similar age in Southwest Japan, the granitoids of central Hokkaido are slightly low in K/Na and in Sn content (about 2.7 ppm) and high in K/Rb (less than 250).

b)   *Outer arc plutonic belt*

In central to southern Kyushu, where the Honshu and Ryukyu arcs intersect, there are few batholithic masses larger than 400 km^2. The

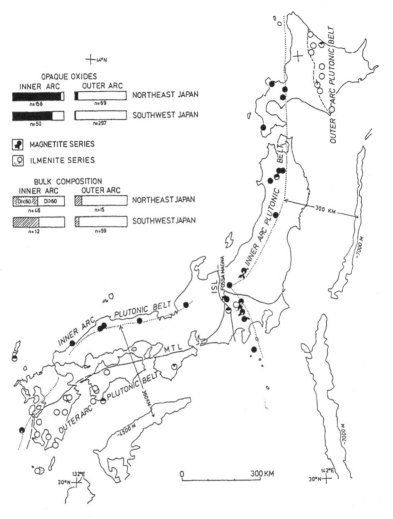

Fig. 4.9. Tertiary plutonic rocks and the ratio magnetite-series rocks/ilmenite-series rocks (after ISHIHARA, 1979a).

representatives are the Okueyama, Minami-Osumi and Yakushima masses, all intruding into the Shimanto belt, although the Omogokei mass in Shikoku is in the Sambagawa belt. In the Okueyama area the upper part of the magma reservoir of the super-caldera is exposed at the present surface (TAKAHASHI, 1986). A few volcanic rock bodies are associated with these granitoids, for instance, the Kumano and the Osuzu volcanic rocks. Further-more, the Mitake monzogranite and the Kaikomagatake granite, both

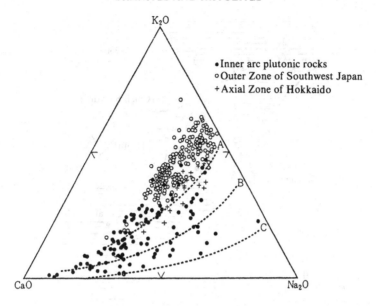

Fig. 4.10.   CaO–K₂O–Na₂O plots of the Neogene plutonic rocks (after ISHIHARA *et al.*, 1976). The broken line A defines the limit of rocks in the Outer Zone of Southwest Japan, and corresponds to the average trend of the Japanese granitoids by ARAMAKI *et al.*, 1972. The Tanzawa-type rocks are plotted in the area between lines B and C.

situated in central Japan, are relatively low in K/Rb (less than 250) and are regarded as masses of the outer belt.

Most granitoids of the outer belt are of the ilmenite-series, but a few of the magnetite-series occur with them in certain areas such as the Ashizuri district, southwest Shikoku. Gigantic (up to 10 cm in length) phenocrysts of potassium feldspar are characteristic of some outer belt granites, the most well-known being those of the Yakushima granites. Furthermore, oxide or silicate minerals rich in aluminum are occasionally found in the outer belt granitic rocks (NOZAWA, 1956). The following geochemical characteristics are observed: $K_2O>Na_2O$, $FeO>CaO$, relatively rich in Rb, Sn and F, high in $\delta^{18}O$ (SiO_2 70%) (+10 to +12‰) and low in $\delta^{34}S$ (−4 to −10‰).

c)   *Inner arc plutonic belt*

The inner belt granitoids comprise intrusives exposed on the inner side of the volcanic front. They are mostly in the Green Tuff Region, and particularly larger masses occur in the southern part of Fossa Magna, central Japan. Granitoids of the Green Tuff region are closely associated with more basic rocks, gabbro or diorite. They mostly lie in the magnetite-series. A somewhat complicated spatial relationship is observed in the Kofu district, where rocks of both series are exposed side by side (Fig. 4.11). The Mitake

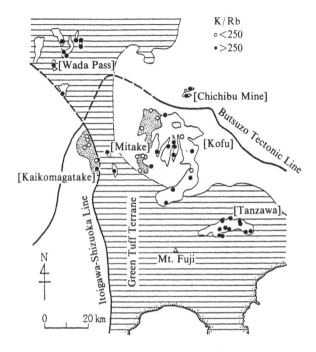

Fig. 4.11. Plutonic rocks and their K/Rb ratios in the southern Fossa Magna area (after Ishihara and Terashima, unpubl.). Stippled: rocks of the Outer Zone type, ( ): name of the mass.

mass consists exclusively of the ilmenite-series rocks; however, the Kaikomagatake mass is comprised of rocks of both series. These masses, therefore, might be regarded as members of the outer belt.

The inner belt granitoids are generally poor in ^{18}O, δ^{18}O (SiO_2 70%) being less than 6‰. δ^{34}S is about +6‰, and the Sr isotope initial ratio is relatively low. The ratio is particularly low, about 0.703, in the Tanzawa tonalite mass. Other geochemical features also suggest that the Tanzawa mass is peculiar among the Japanese granitoids and seems to be of the M-type.

4.4 Summary

As stated in the foregoing sections, distinctive zonal distribution of the magnetite-series and the ilmenite-series granitoids are observed in the Japanese Islands. Considering the Neogene granitic rocks, rocks of the former series appear in the inner belt, which roughly coincides with the area of Quaternary volcanoes. The inner belt granitoids are, therefore, considered to have been formed in relation to the westward subduction of the Pacific and the Philippine Sea plates under the Japanese Islands. On the other hand, no obvious relation is seen between the distribution of outer belt granitoids and the subduction zone. The parental material of the magnetite-series rocks would have been generated in the upper mantle overlying the subducting plate, but those of the ilmenite-series rocks are considered to have been derived from thick sediments deposited in the fore-arc region.

A similar spatial relationship is observed in the Cretaceous granitoids. The magnetite-series rocks increase in quantity towards the inner marginal sea areas, although in places the relationship is not sufficiently distinctive.

REFERENCES

ARAMAKI, S., K. HIRAYAMA, and T. NOZAWA, Chemical composition of Japanese granites, Part II: Variation trends and average composition of 1200 analyses, *J. Geol. Soc. Japan*, **78**, 39049, 1972.

ISHIHARA, S., The magnetite-series and ilmenaite-series granitic rocks, *Mining Geology*, **27**, 293–305, 1977.

ISHIHARA, S., Metallogenesis in the Japanese island arc system, *J. Geol. Soc. London*, **135**, 389–406, 1978.

ISHIHARA, S., Lateral variation of magnetic susceptibility of the Japanese granitoids, *J. Geol. Soc. Japan*, **85**, 1979.

ISHIHARA, S., H. KANEYA, and S. TERASHIMA, Genesis of the Neogene granitoids in the Fossa Magna region in Japan, *Marine Sci. Month.*, **8**, 523–528, 1976.

ISHIHARA, S. and S. TERASHIMA, Chemical variation of the Cretaceous granitoids across southwestern Japan: Shirakawa-Toki-Okazaki transection, *J. Geol. Soc. Japan*, **83**, 1–18, 1977a.

ISHIHARA, S. and S. TERASHIMA, The Tin content of the Japanese granitoids and its significance on the Cretaceous magmatism, *J. Geol. Soc. Japan*, **83**, 657–664, 1977b.

KATADA, M., M. YOSHII, S. ISHIHARA, Y. SUZUKI, C. ONO, T. SOYA, and H. KANAYA, Cretaceous granitic rocks in the Kitakami mountains-Petrography and zonal arrangement, *Geol. Surv. Japan Rept.*, **251**, 1974.

KAWANO, Y. and Y. UEDA, K-A dating on the igneous rocks in Japan (II)-Granitic rocks in Kitakami massif, *J. Japan. Assoc. Mineral. Petrol. & Econ. Geol.*, **53**, 143–154, 1965.

MATSUMOTO, T., Timing of geological events in the circum-Pacific region, *Can. J. Earth Sci.*, **14**, 551–561, 1977.

MIMURA, K., M. KATADA, and H. KANAYA, Igneous activity of the Kyoto rhyolites in the Yatsuoyama district, southeast of Lake Biwa, *J. Japan. Assoc. Mineral. Petrol. & Econ. Geol.*, **71**, 327–338, 1976.

MURAKAMI, N., Two contrastive trends of evolution of biotite in granitic rocks, *J. Japan. Assoc. Mineral. Petrol & Econ. Geol.*, **62**, 223–248, 1977.

MURAKAMI, N., A consideration on the mechanism of formation of the Paleogene Tamagawa cauldron, southwest Japan, *Mem. Geol. Soc. Japan*, **9**, 93–105, 1973.

NOZAWA, T., Some considerations on the spinel-bearing inclusions in Osuzu acid rock, Kyushu, Japan, *J. Japan. Assoc. Mineral. Petrol.*, **40**, 33–38, 1956.

RESEARCH GROUP FOR LATE MESOZOIC IGNEOUS ACTIVITY OF SOUTHWEST JAPAN, Late Mesozoic igneous activity and tectonic history in the Inner Zone of Southwest Japan, *Assoc. Geol. Collab. Japan. Monogr.*, **13**, 1967.

SASAKI, A. and S. ISHIHARA, Sulfur isotopic composition of the magnetite-series and ilmenite series granitoids in Japan, *Contrib. Mineral. Petrol*, **68**, 105–117, 1979.

SHIBATA, K. and S. ISHIHARA, K-Ar ages of biotites across the central part of the Hiroshima granite, *J. Geol. Soc. Japan*, **80**, 431–433, 1974.

SHIBATA, K. and S. ISHIHARA, Initial ^{87}Sr/^{86}Sr ratios of plutonic rocks from Japan, *Contrib. Mineral. Petrol*, **70**, 381–301, 1979.

TAKAHASHI, M., Anatomy of a middle Miocene Valles-type caldera cluster: Geology of the Okueyama volcano-plutonic complex, Southwest Japan, *J. Volcanol. Geotherm. Res.*, **29**, 33–70, 1986.

TANAKA, K. and T. NOZAWA (eds.), *Geology and Mineral Resources of Japan, Vol. 1: Geology*, Geol. Surv. Japan, 1977.

YAMADA, N., T. NOZAWA, Y. HAYAMA, and T. YAMADA, Mesozoic felsic igneous activity and related metamorphism in central Japan: From Nagoya to Toyama, *Guidebook for Excursion 4*, Geol. Surv. Japan, 1977.

CHAPTER 5

Chapter 5

LATE CENOZOIC STRATA

A distinct discontinuity in the geology of the Japanese Islands exists at the boundary between the Paleogene and Neogene to Quaternary. A great subsidence commenced during the Middle Miocene, particularly in Northeast Japan, and this was accompanied by immense volcanism. Volcanic rocks of this age are frequently altered into pale green rocks, called "Green Tuff" by Japanese geologists. The area where they are most widely developed is named the "Green Tuff Region". Subsequently the sedimentary basins in the Sea of Japan region of Northeast Japan became shallower, and a major upheaval occurred in Quaternary times. The late Cenozoic strata of this region as a whole represent a major cycle of sedimentation, that commenced with subsidence and ended with upheaval. On the other hand, in middle to western Hokkaido, the outer belt of Southwest Japan and a part of Kyushu, sedimentation continued through to Neogene times, with or without intermission, from the close of Paleogene times.

In this chapter Neogene to early Quaternary strata are described. They are grouped into 10 provinces, from Hokkaido to the Nansei-shoto archipelago.

The paleontological and biostratigraphical studies of the Japanese late Cenozoic are mentioned in the following publications.

> IUGS Commission on Stratigraphy, Regional Committee on Stratigraphy (ed.), Proceedings of the 1st International Congress on Pacific Neogene Stratigraphy, Kaiyo Shuppan Co., 1977
>
> Huzita, K. *et al.* (eds.), Cenozoic Geology of Japan, Department of Geology, Osaka City University, 1978
>
> Tsuchi, R. (ed.), Fundamental Data on Japanese Neogene Bio-and Chrono-stratigraphy, Shizuoka University, 1979

5.1 Hokkaido

The Neogene terranes of Hokkaido are divided into five sub-terranes as shown in Fig. 5.1 and classified in Table 5.1.

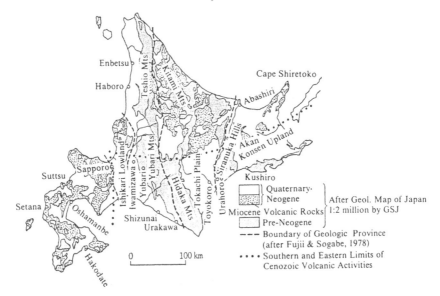

Fig. 5.1.   Hokkaido.

Table 5.1.   Correlation of the principal Neogene strata of Hokkaido. G: Group.

Age		Plank. Foram. Z	Oshima Peninsula	Ishikari	Rumoi	Shizunai[1]	Kushiro Coal Field	Tsubetsu[4]	Kitami Mts.[5]
Quat.		N23	Setana*				Tokachi G	Akan G	
		N22	Kuroma-tsunai	Takigawa	Mochiku-betsu				
Pliocene		N21	Tate*				Oikamanae G	Kamisato G	
		N19							
		N18		Oiwake	Enbetsu				
	Late		Yakumo						
Miocene		N17		Iwami-zawa	Kinkoma-nai				
		N16							
		N15							
		N14							
	Middle	N10			Kotan-betsu	Ukekoi	Okawa G	Tsubetsu G	Horoka
		N9	Kunnui•	Kawabata					
		N8		Takinoue	Furanui				
			Yoshioka						
		N7			Haboro				
	Early	N6	Fukuyama				Touberi G		

*Tsuchi (1979), 1) Matsuno-Yamaguchi (1958), 2) Miyashaka-Kikuchi (1978), 3) Mizuno et al. (1969), 4) Yamaguchi (1966), 5) Yamagishi (1976)

5.1.1 Easternmost Hokkaido

The area to the east of the Abashiri-Urahoro line (line a in Fig. 5.1) is the geological extension of the Kurile arc, where the inner volcanic and the outer non-volcanic zones are distinguished. In easternmost Hokkaido, an uplift zone extends in a NE-SW direction from the Shiretoko Peninsula to the Shiranuka Mountains. In the former area the Neogene strata constitute the basement on which Quaternary volcanoes lie, while in the latter area the Cretaceous to Paleogene rocks are exposed. The uplifted terrane is thrust up to the west over the Kitami-Tokachi zone, suggesting, in conjunction with the tectonic structures observed in the Tokachi and Hidaka areas to the west, that the western part of the Kurile arc collides with western Hokkaido (FUJII and SOGABE, 1978).

5.1.2 Kitami-Tokachi zone

The Neogene strata overlying the pre-Cenozoic rocks in the northern part, the Kitami Mountains, of this zone mostly consist of altered subaerial pyroclastic rocks, which are occasionally associated with ore deposits of various metals. On the other hand, the Tokachi district in the southern part is underlain by thick piles of Upper Cenozoic sediments, ranging from early Miocene to Quaternary. In the lowermost part of the strata, subaerial volcanic rocks (about 15 Ma) of alkalic rhyolite composition occur together with a few coal seams. However, Middle to Upper Miocene consists of thick mudstone beds with some volcanic materials, and the upper part contains abundant coarse clasts of metamorphic and plutonic rocks of the Hidaka belt to the west (MIYASAKA and KIKUCHI, 1978).

The Plio-Pleistocene rocks of this zone are collectively called the Tokachi Group (TOKACHI RESEARCH GROUP, 1978), whose sedimentary basin migrated from east to west as time went on.

5.1.3 Teshio-Ishikari zone

Thick Neogene sedimentary piles overlie Cretaceous to Paleogene rocks in the Teshio-Ishikari zone to the west of the axial mountains of Hokkaido. The deposits as a whole represent one cycle of sedimentation, which commences with non-marine coal-bearing sediments, grading into marine sandstone and mudstone and finally closing with shallow marine sediments. The Pliocene sediments of the Haboro area to the north contain shells indicative of a cold current, but shells of the Yubari area to the south indicate a warm current (CHINZEI, 1978).

The Neogene strata of the Teshio-Ishikari zone are folded with many axes stretching in the N-S direction. The folds in the Yubari area are in many cases asymmetrical with steeper west wings and gentler east wings, and they are frequently associated with east-dipping thrusts. In the Ishikari Lowland to the west the structures become more moderate (Fig. 5.2). The structures would have resulted from westward upthrusting of the axial mountains lying to the east.

5.1.4 Southwestern Hokkaido

To the east of the Kuromatsunai Lowland running from Shuttsu to Oshamambe, no exposures of pre-Cenozoic rocks are found, and abundant Quaternary volcanic rocks overlie the Neogene strata. To the west of the lowland, however, a few higher mountaineous areas are underlain by Mesozoic sedimentary rocks and Cretaceous granites, and here the Neogene rocks are distributed surrounding these higher areas. Neogene sedimentation commenced with subaerial volcanism, however sedimentation subsequently continued in a marine environment, still with the addition of abundant volcanic materials.

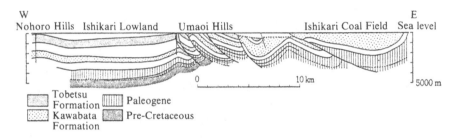

Fig. 5.2.   Cross-section of the Ishikari lowland and the Ishikari coal-field (after MITANI, 1978).

5.2 Tohoku Province

The axial mountain range (called the Backbone Range), 1000 to 1500 meters high, runs in a N-S direction in Tohoku province, Northeast Japan (Fig. 5.3). To the west of this range, three zones run parallel to each other, also in a N-S direction. They are, from west to east, the Sea of Japan coastal plains, the Dewa Hills and the intermontane basins. On the other hand, the Kitakami and Abukuma Mountains are located to the east of the axial range, the Kitakamigawa Lowland lying between them (Fig. 5.3).

The Sea of Japan coastal plains are underlain by thick piles of Neogene sedimentary and volcanic rocks. The area as a whole represents the widest of the Japanese Neogene terranes. The pre-Cenozoic basement rocks are exposed sporadically in the mountainous areas, for example, in the Shiragami and the Budo districts and pre-Cenozoic basement is overlain by the Lower Miocene in the marginal part of the mountains. The Upper Miocene to

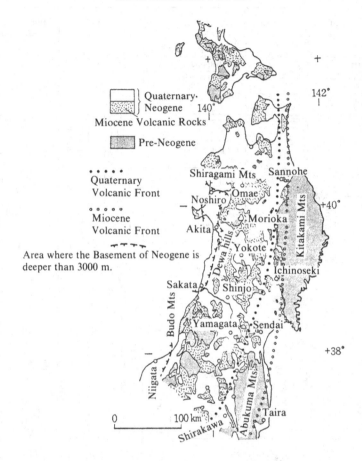

Fig. 5.3. Tohoku province (based on the 1:2,000,000 GEOLOGICAL MAP OF JAPAN issued by Geological Survey of Japan, 1978).

Pliocene deposits, on the other hand, mostly occur in the Dewa Hills, the intermontane basins and the coastal plains.

The Neogene to Quaternary strata of the Akita Plain are stratigraphically classified as shown in Fig. 5.4. They are folded, with axes running roughly in a N-S direction; at the crests of anticlines petroleum may occur (FUJIOKA et al., 1977).

The Oga Peninsula is situated to the northwest of the Akita Plain. The lowermost Neogene Nishioga Group are exposed in the westernmost part of the peninsula. It is successively overlain by younger strata: Daijima, Nishikurosawa, Onnagawa, Funakawa, Kitaura and Wakimoto Formations, which all dip to the east. The lower part of the Kitaura Formation is roughly

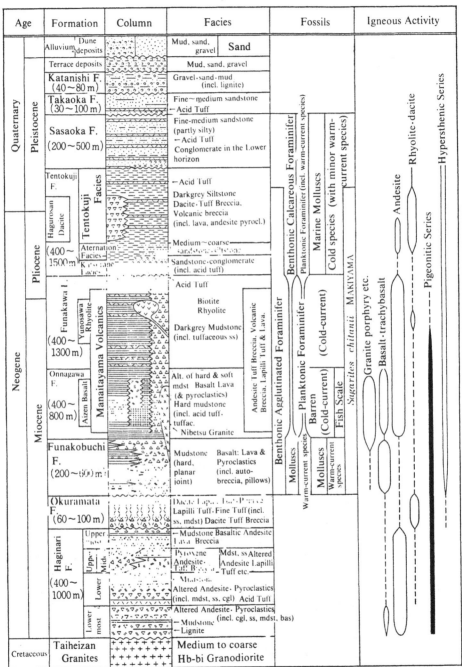

Fig. 5.4. Compiled columnar-section of Neogene strata in the Akita district (modified after FUJIOKA *et al.*, 1977). Geological age by TSUCHI (1979). See Table 5.2.

correlated with the Tentokuji Formation in the Akita Plain (KITAZATO, 1975).

Further to the south of the Akita Plain, the folded and oil-bearing Neogene strata are exposed in the Kitayuri Hills and the Shonai Plain. The folded structure is asymmetrical with a steeper west wing. An eastward-dipping reverse fault runs along the western margin of the Kitayuri Hills; its stratigraphic throw reaches 2000 m (Fig. 5.5).

The Dewa Hills are mostly underlain by thick and fairly uniform mudstone beds, ranging from Middle to Upper Miocene (TAGUCHI, 1962). The strata are gently folded, and the folding movement has continued up to Recent times, resulting in warping of river terraces and vertical movement of bench marks.

Between the Dewa Hills and the Backbone Range few intermontane basins lie in a N-S direction. They are, from north to south, the Hanawa, Yokote, Shinjo and Yamagata Basins. They are separated from each other by the uplifted mountainous regions of Miocene volcanics. These basins are underlain by Middle to Upper Miocene marine formations (TAGUCHI, 1962). Most of the Pliocene strata are of fresh-water origin and include coal-seams. Some of the strata of the intermontane basins are gently folded, and movements are still active today, distorting fluvial terrace surfaces (SUGIMURA, 1967). Active Quaternary faults sometimes define the peripheries of the basins and these have sometimes caused earthquakes: for instance, the Rikuu Earthquake of 1896.

The early Miocene strata (Oarasawa and Oishi Formations) of the Backbone Range consist of marine sedimentary rocks associated with volcanics. This is in contrast to the continental nature of the Daijima Formation in the Oga Peninsula, which is of similar age. The strata are overlain by tuffaceous marine sandstone and mudstone (the Koshigesawa and Kurosawa Formations) of middle to late Miocene age (Fig. 5.5). Finally, the Upper Miocene strata comprise subaerial dacitic pyroclastic flow deposits and terrestrial sediments.

The Lower to Middle Miocene of the lowland between the Backbone Range and the Kitakami Mountains are exposed in a few separated areas such as the Sannohe, Ichinoseki, Sendai and Fukushima areas (CHINZEI, 1966). The lower part consists of volcanic rocks covered by Middle Miocene warm-sea sediments and further by fine deeper sea sediments. The facial change is comparable with that observed in areas to the west of the Backbone Range. The Neogene strata of the lowlands are generally folded and tilted eastwards with generally gentle, but sometimes steep, dips (e.g. MATSUNO, 1967).

Along the eastern border of the lowland, early Miocene volcanic rocks are exposed, which exhibit andesitic to dacitic compositions. Further south,

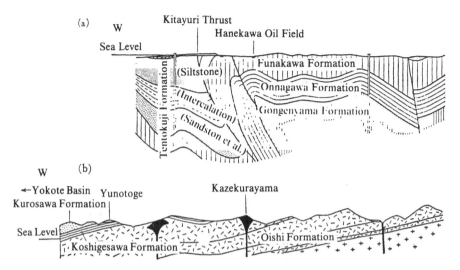

Fig. 5.5. Cross-section of the Sea of Japan coastal areas and the axial range of the Tohoku province. No vertical exaggeration. (a) Kitayuri Hills, Akita Prefecture (after FUJIOKA *et al.*, 1976), (b) Wakagawa area, Iwate Prefecture (after OSAWA *et al.*, 1971).

Miocene volcanic and pyroclastic rocks occur sporadically in the Kujigawa and Joban areas. It is suggested that the volcanic front of Miocene times was located 20 to 60 km to the east of the present-day front.

No Neogene stratum is exposed in the Kitakami and Abukuma Mountains, which separate the Neogene terranes to the west and the basins in Pacific Ocean off Northeast Japan.

Fault scarps up to 400 meters high are developed along the Futaba Fault (TSUNEISHI, 1978) which runs parallel to the eastern border of the Abukuma Plateau. In the area to the east of the fault scarps the Neogene marine beds are exposed. The strata extend southwards to the Joban area, where they overlie the Paleogene coal-bearing sediments. Miocene strata underlie small areas forming a narrow belt along the Tanakura shear zone (OTSUKI, 1975).

5.3 Shin-etsu Region

In the area extending along the Sea of Japan coast of central Japan, from north of Niigata City to Itoigawa, the Niigata oil field occurs, it is underlain by Miocene and younger strata. An extension from Itoigawa southwards to Matsumoto, is called the northern Fossa Magna region (Fig. 5.6).

The Neogene to Quaternary strata of the Niigata oil field are very thick

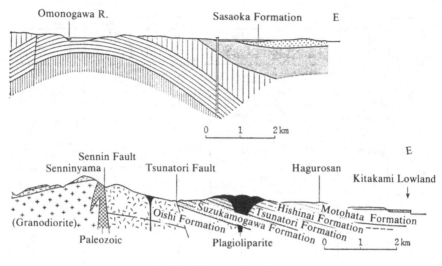

Fig. 5.5. (continued.)

and reach 5000 to 6000 m in depth in the central part of the field (KATAHIRA, 1974; IKEBE *et al.*, 1978). The lowermost part consists of volcanic rocks and shale, and they grade upwards into slightly deeper marine sediments and then finally into shallow-sea~brackish~non-marine clastic sediments. The thick sedimentary piles are divided stratigraphically, from lower to upper, into the Nanatani (hard shale, volcanic rocks, middle Miocene), Teradomari (mainly mudstone), Shiiya (intercalated sandstone-mudstone, late Miocene), Nishiyama (massive mudstone), Haizume (shallow-sea sandy silt) and Uonuma (bay to terrestrial sand-gravel-clay, Quaternary) Formations.

The strata are all folded with axes running en echelon in a NNE-SSW direction. The hilly areas correspond to the anticlines, while the basinal areas correspond to the synclines. It is suggested that the folding is still active today. The warping of the Shinanogawa river terraces in the Sekihara-Katagai and Ojiya districts can be observed even by the unaided eye (OTA and SUZUKI, 1979).

Westward dipping reverse faults (Shinanogawa fault zone) are developed in the northeastern peripheries of the Niigata oil field. They are still active and caused the Zenkoji earthquake in 1847. The 1964 Niigata earthquake was also the result of reverse faulting occurring along the N-E extension of this fault zone.

High mountainous regions such as the Asahi, Iide and Echigo Mountains

Table 5.2. Correlation of the principal Neogene strata of Tohoku province.

Age	Ma	Plank. Foram. Z.	Sea of Japan Coast					Backbone Range		Eastern Marginal Area			
			Oga	Akita	Niigata	Aizu**	Nagano	Wagagawa	Shinjo***	Sendai	Ichinoseki	Sannohe	Joban
Quat.	0	N23	Wakimoto*	Sasaoka*	Haizume*	Nanaorizaka			Yamaya	Dainenji*			
		N22	K'*'*'*				Sarumaru			Mukaiyama*			
Plioc.		N19	Funakawa*		Nishiyama*	Izumi	Saku		Izumikawa Shimizu Yamuke Sakekawa	Tatsunokuchi* Kameoka*		Sannohe G.	Taga*
	5	N18		Funakawa*	Shiiya*	Fujitoge	Ogawa			Mitaki* ---		Kubo*	
		N17	(Mayama Diatom)*		Teradomari*	Shiotsubo	Aoki	Hanayama	Noguchi Furukuchi	Shirasawa Tsunaki		Shitazaki*	
Late	10	N16						Kurosawa				Tomesaki*	
		N15	Onnagawa*	Onnagawa*		Urushikubo	Bessho	Kotsunagizawa	Kusanagi	Hatatake*	Shimokurosawa*		
Middle		N14				Ogino							
		N10	Nishikurosawa*		Nanatani*	Kagata	Uchimura				Suenomatsuyama*	Shimotakaku*	
Miocene	15	N9		Sunakobuchi				Oishi		Moniwa*	Kadonosawa*	Numanouchi*	Yunagaya G.
		N8	Daijima	Okuramata			Moriya			Takadate		Yotsuyaku*	
		N7								Tsukimoki			
Early	20	N6	Nishioga	Haginari				Oarasawa	Nozoki				

*Tsuchi (1979), **Suzuki et al. (1977), ***Taguchi (1962), Nakagawa et al. (1971), —·— Ikebe (1978).

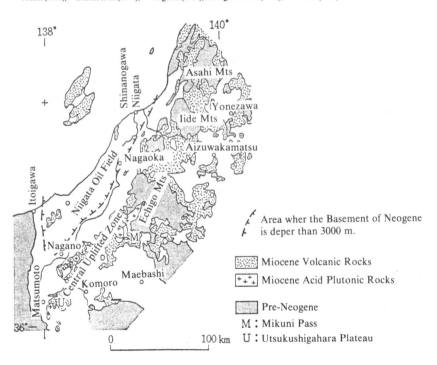

Fig. 5.6. Shin-etsu region (based on the 1:2,000,000 Geological map of Japan).

are underlain by pre-Cenozoic sedimentary and granitic rocks. The inter-montane areas have exposures of Miocene marine sediments and volcanic rocks. Terrestrial sediments are widely exposed in the Aizu Basin area, suggesting the development of lakes in Quaternary time. The sediments are frequently intercalated with acidic tuffs as exemplified by the Shirakawa Acidic Pyroclastics (SUZUKI et al., 1977).

The thickness of the Neogene strata is drastically changed at the western margin (Matsumoto-Nagano line) of the central zone of uplift in the northern Fossa Magna region. To the west of the line, the Neogene marine sediments are very thick and classified stratigraphically, from lower to upper horizons, into the following five formations: Bessho (black mudstone, middle Miocene), Aoki (sandstone-mudstone), Ogawa (intercalation of conglomerate, sandstone and mudstone, late Miocene), Shigarami (andesite, conglomerate-mudstone, Pliocene), and Sarumaru (sandstone-conglomerate, acidic tuff, Pliocene to Holocene). The last is further overlain by the lacustrine Toyono Formation of middle Quaternary (TOYONOSO RESEARCH GROUP, 1977).

On the other hand, the central uplifted zone to the east is underlain by Lower to Middle Miocene volcanic and marine sedimentary rocks which are in some places intruded by quartz diorite. The strata are divided, in the ascending order, into the Moriya, Uchimura, Bessho and Aoki Formations. Therefore, the Neogene of this area includes lower strata than those of the western areas described above. In the Ueda-Komoro district, further to the east of the uplifted zone, strata lower than the Ogawa Formation are not developed, and in this locality the Middle Miocene is covered directly by the Plio-Pleistocene lacustrine Komoro Group.

5.4 Kanto Plain

The Kanto Plain, the largest plain in Japan, was developed as a great sedimentary basin in late Cenozoic times.

Mountainous older rock terranes surround the plain on the western, northern and southern sides and become lower in elevation towards the central part of the plain where they are hidden under the thick late Cenozoic sedimentary pile. The base of the sedimentary pile lies at about 3000 m depth in the center of the plain (KAKIMI et al., 1973). The lower part (about 2000 m) of the pile is of Miocene age and consists of clastic marine sediments with intercalated tuffs, and the upper part (less than 1000 m) is sediments of the marine Plio-Pleistocene Kazusa Group. The Miocene is exposed on the western periphery of the plain and thins eastwards and northeastwards (Fig. 5.7). In the lower Tone River area the pre-Miocene basement becomes abruptly shallow and is overlain directly by the Kazusa Group, the Miocene

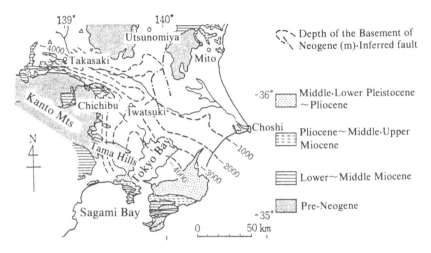

Fig. 5.7. Kanto province (based on the 1:2,000,000 Geological Map of Japan and on KAKIMI *et al.*, 1973).

being lacking. The Kazusa Group is particularly widely exposed in the Boso and Miura Peninsulas and the Tama and Hiki Hills.

In the Hayama-Mineoka belt crossing the two peninsulas in an E-W direction the pre-Neogene Mineoka Group is exposed as well as the Lower Miocene Hota Group and the Middle Miocene Sakuma Group which lie on the Mineoka Group. Serpentinite, gabbro and pillowy basalt are associated with the latter.

Marine sedimentary piles of Middle Miocene to Late Quaternary age are widely exposed in the central to northern part of the Boso Peninsula. The strata are gently tilted northwards. Consequently, the younger beds are successively exposed in the northern areas. The older strata range from the Miocene Amatsu Formation to the Pliocene Anno Formation, and show an E-W strike; however, the Plio-Pleistocene Kazusa Group strikes N-E. A distinctive structural discordance exists between them, known as the Kurotaki unconformity (KAWABE *et al.*, 1979).

The Late Cenozoic strata of the Boso Peninsula exhibit relatively simple structures and few stratigraphic gaps with the exception of the Kurotaki unconformity. Consequently, detailed studies on stratigraphy, micro-fossils and paleomagnetism have been possible. The results are summarized in Table 5.3. In the Miura Peninsula to the west, the Miura Group unconformably overlies the Lower Miocene Hayama Group of the Hayama-Mineoka belt.

To the south of the belt, Miocene marine strata are also distributed.

Table 5.3. Correlation of the Neogene strata of southern Kanto province, and the relations between principal tephra horizons, magnetic stratigraphy and geological age. G: Group.

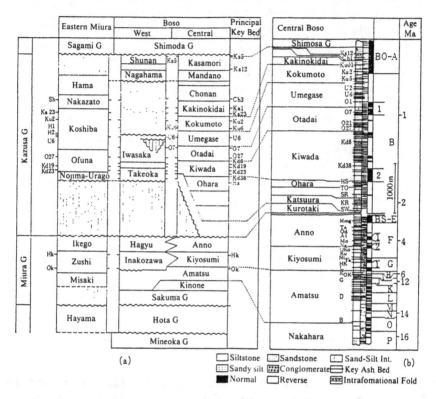

(a)

☐ Siltstone ☐ Sandstone ☐ Sand-Silt Int. (b)
⬚ Sandy silt ⬚ Conglomerate ⬚ Key Ash Bed
■ Normal ☐ Reverse ⬚ Intrafomational Fold

They are, however, strongly folded with northward dips and are frequently cut by reverse faults. These highly deformed strata are considered to be the eastern extension of the southern Fossa Magna region described below.

In the eastern foot hills area of the Kanto Mountains (e.g. Hiki Hills) and further into the mountainous area (e.g. Chichibu Basin) Miocene marine strata out-crop.

5.5 Southern Fossa Magna

A reversed V-shaped bending of the zonal structure of the Outer Zone of Southwest Japan is observed in central Japan, on the Pacific side of which is located the southern Fossa Magna region (Fig. 5.8). No pre-Cenozoic rock is exposed there, and the region is underlain by extremely thick piles of marine sediments and volcanic deposits of Miocene and later age. Regarding volcanic activity, this area belong to the inner zone (volcanic zone) of

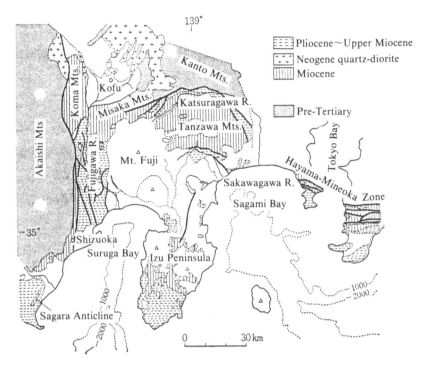

Fig. 5.8.   Southern Fossa Magna area (after KAKIMI *et al.*, 1973).

northeast Japan, however, the nature of the sedimentary basins and the Neogene structures are comparable with those of the outer zone (ocean-side zone) of Southwest Japan. It is said, therefore, that the geological features of the inner zone of Northeast Japan and the outer zone of Southwest Japan overlap each other in the southern Fossa Magna region.

The Early to Middle Miocene strata of the southern Fossa Magna consist mainly of submarine volcanic materials and associated mudstone and are exposed in the Tanzawa, Misaka and Koma Mountains from east to west (MIKAMI, 1961; KOSAKA and TSUNODA, 1969). These mountains represent the anticlinorium zone uplifted since the late Miocene.

The Late Miocene and younger marine sediments associated with andesitic volcanic rocks are exposed in subsiding zones fringing the uplifted mountains. They are termed as follows in each area located clockwise from west to east: Fujigawa (MATSUDA, 1961), Nishikatsura, Aikawa (MIKAMI, 1961) and Ashigara Groups. Particularly in the upper part of the formations turbidites are developed containing abundant pebbles and boulders derived from the surrounding Kanto and Akaishi Mountains and from the uplifted mountains within the southern Fossa Magna. For instance, the lower part of the Ashigara Group carries pebbles of pre-Cenozoic shale and sandstone of the Kanto Mountains, and the upper part contains boulders of Miocene volcanics, crystalline schists and plutonic rocks of the Tanzawa Mountains. The change in lithology of the conglomerates suggests an uplifting history for the mountains and the synchronous development of the subsiding belt (MATSUDA, 1962).

The Neogene strata of southern Fossa Magna are strongly folded to dip steeply, and are displaced by a number of reverse faults with west- or northward dips. The Eocene to Early Miocene rocks of the Shimanto belt are also thrust up over the Fossa Magna Neogene. Even the late Quaternary sediments and ejecta of Mt. Fuji volcano are disrupted by these faults.

In the Tanzawa Mountains the Tanzawa Group makes an east-west elongated domal structure, whose central part is occupied by a quartz diorite mass. Metamorphic rocks are particularly developed on the southern side, the grade varying from the prehnite-pumpellyite, through the greenschist to the amphibolite facies. The zeolite facies recrystallization took place later over the whole metamorphic area (SEKI et al., 1969). The Kannawa and Kozu-Matsuda faults run along the southern border of the mountains. They extend to the inferred plate boundary in the Sagami Trough. The 1923 Great Kanto Earthquake is considered to have occurred as a result of the reverse right-lateral fault movement at the boundary.

Izu Peninsula projects southwards and divides the Nankai Trough into two parts: the Suruga Trough to the west and the Sagami Trough to the east. The peninsular geology is remarkably different from that of the other Fossa Magna areas. It is comprised of a number of volcanic mountains formed since Miocene times. The lowermost rocks are the Nishina Group of Early Miocene age, and they are covered by the Yugashima and Shirahama Groups of Middle Miocene to Pliocene age, both include abundant volcanic rocks (KITAMURA et al., 1969). Furthermore, a number of Quaternary volcanoes have been developed in the peninsular and contiguous districts, being represented by the Amagi Volcano (KURASAWA, 1959).

5.6   Hokuriku and San-in

The Sea of Japan coastal area of Southwest Japan is considered geologically to be a part of the Green Tuff Region (Fig. 5.9 and Table 5.4) (FUJITA et al., 1978). The principal Neogene terranes are distributed in the Hokuriku area extending from Toyama to Fukui and in the San-in area to the west of the Tango Peninsula.

In the Hokuriku area a few plains are distributed such as the Toyama, Tonami, Kaga and Fukui Plains, each separated by hilly districts stretching

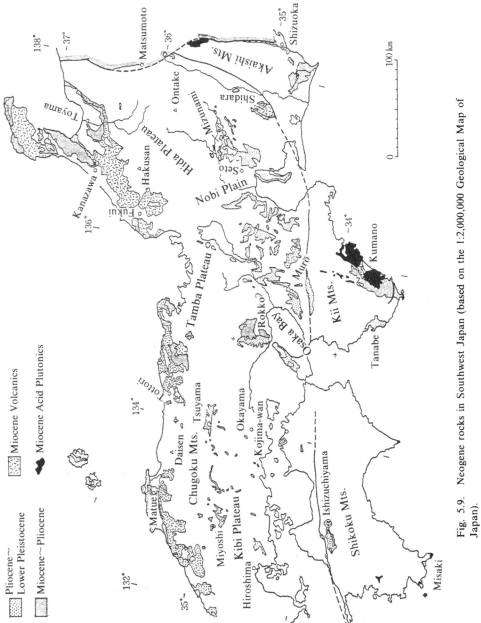

Fig. 5.9. Neogene rocks in Southwest Japan (based on the 1:2,000,000 Geological Map of Japan).

Table 5.4. Correlation of the principal Neogene strata of the northern Kyushu, San-in, Hokuriku and Setouchi areas. G: group.

Age	Ma	Plank. Foram. Z.	Kitakyushu	San-in Yamaguchi	Shimane	Hokutan	Hokuriku Kanazawa	Toyama	Chugoku	Kinki	Mizunami
Quat.	0 — 2	N23 / N22	Oita G				Omma*			Osaka G	Seto G
Plioc.		N21 / N19						Yabuta*			
Miocene Late	5	N18	Kitamatsu Basalt			Teragi G	Takakubo	Otokawa*			
		N17			Matsue						
	10	N16			Fujina		Shimoaraya				
		N15 / N14			Omori						
Miocene Middle	15	N10 / N9	Nojima G / Sasebo G	Yuyawan G	Kuri*	Muraoka / Toyooka	Asagaya / Nanamagari / Sunakosaka Kurosedani*	Higashibessho*	Bihoku G*	Kobe G / Nijo G	Mizunami G*
		N8 / N7			Kawai* / Hata	Yoka	Iozen	Iwaine			
Miocene Early	20	N6	Nishisonogi G / Matsushima G	Hioki G		Takayanagi					

*Tsuchi (1979), —·— Ikebe (1978)

in a NNE-SSW direction (SAKAMOTO, 1966; KASENO, 1977; FUJII et al., 1970). The Neogene strata are mostly exposed in these hilly districts, and the plains are underlain by the Quaternary. The plains and the hills are bounded by zones of flexure and fault.

The Neogene strata of the Hokuriku area are collectively called the Hokuriku Group and consist, for example in the Toyama district, of the following five formations, from lower to upper: 1) basal conglomerate (Nirehara Formation), 2) andesite (Iwaine Formation and others) and rhyolite (Iousen Formation), 3) marine sediments with subtropical shells (Yatsuo Formation), 4) marine sandstone and mudstone (Otokawa Formation, Himi Formation and other), and 5) fresh-water clastics (Hanyu Formation). They represent a cycle of sedimentation.

In the central part of the Noto Peninsula, projecting northwards into the Sea of Japan, is the Ouchi Lowland. This is a compressional graben and extends southwestwards to the eastern part of the Kaga Plain. The area to the north of the lowland is underlain by Miocene andesitic rocks overlying plutonic rocks of the Hida metamorphic terrane.

In the hilly areas to the west of the Fukui Plain late Miocene strata are exposed in the northern, early Miocene beds in the middle and older

basement rocks in the southern area. A group of faults runs along the eastern
border of the Fukui Plain. The Fukui Earthquake of 1948 was caused by one
of these faults.

Wide coastal plains have not been developed in the San-in district west
of the Tango Peninsula. Neogene strata are only exposed in narrow coastal
areas. But in some places they extend southwards, making a few embayments
such as those of the Chiyogawa (UEMURA *et al.*, 1979), Matsue (SAN-IN
GREEN TUFF RESEARCH GROUP, 1979), Izumo and Ota. Furthermore, a few
isolated Neogene terranes occur in the Chugoku Mountains of pre-Cenozoic
rocks.

A remarkable feature of the Neogene in the San-in district is that
terrestrial sediments and subaerial volcanic rocks are abundant in the lower
horizons (Middle Miocene); however, the upper part consists of normal
marine mudstone. An example of this is the Hokutan Group in eastern
San-in, whose lower part is composed of non-marine basal conglomerate and
subaerial volcanic rocks (e.g. Yoka Formation), but the upper part consists
of marine mudstones (Muraoka Formation). The latter is further overlain by
lacustrine sediments and volcanic rocks of Late Miocene to Pliocene age (e.g.
Teragi Formation). The Pliocene lacustrine sediments carry, in some places,
sedimentary uranium ore deposits (FUJITA, 1972).

The Neogene strata of the Shimane Peninsula form an anticlinorium on
the north of the Shinji-Nakanoumi synclinorium (TAI, 1973) in the mid
San-in region.

5.7 Setouchi

The Setouchi Province comprises an area on the northern side of the
Median Tectonic Line, which extends from the Seto Inland Sea area
eastwards through the central Kinki region to the Mizunami and Shidara
districts of central Japan.

In the Chugoku region Miocene marine strata (of the First Setouchi
Stage) are sporadically exposed in the crestal areas (about 1000 m in altitude)
of the Chugoku Mountains (TAI, 1975), in the intermontane basins such as
the Tsuyama, Niimi-Shobara and Miyoshi districts (all 200–400 m high), and
on the Kibi Plateau (400–600 m high) to the south. Further to the south,
Miocene marine beds have been cored off-shore at 375–70 m depth under
Kojima Bay of the Seto Inland Sea. Thus the base of the Miocene strata in
this region varies in elevation in harmony with the relief of the Chugoku
Mountains, and this variation seems to have resulted mostly from the
Quaternary crustal movement. The lower part of the Miocene consists of
sandstone and conglomerate containing shells of shallow and warm seas
(ITOIGAWA and SHIBATA, 1973). Mudstones containing planktonic foraminifera

are the principal sediment type in the middle part, suggesting that the inland sea of that time was connected with the open sea (TAI, 1959). The strata are collectively called the Bihoku Group.

In the Kinki region similar Miocene strata are distributed in several isolated districts and termed separately, for example, the Fujisawa Group of the Nara district. Further to the east, Miocene marine sediments similar to the Bihoku Group are exposed in the Mizunami area (UEMURA, 1961), in the Ina valley district (UI, 1970), in the Shidara Basin (SHIDARA RESEARCH GROUP, 1979) and in the Chita Peninsula.

After the First Setouchi Stage, volcanic activity (10–15 Ma) occurred in a few districts separated at about 100 km interval. The representative areas are the central Kyushu, Sanuki, Shodo-shima, Nijosan and Shidara districts. Rocks erupted are andesite to dacite, relatively rich in K_2O (NAKADA and TAKAHASHI, 1979).

The Second Setouchi Stage commenced several million years after the volcanic activities ceased. The depositional basins of this stage were limited to the areas just surrounding the present-day inland sea (HUZITA, 1962). The strata representing this stage are the Osaka Group (ITIHARA *et al.*, 1975) of the Osaka Bay area (Fig. 5.10 and 5.11). The lower part is terrestrial, while the upper part contains abundant intercalated beds of marine clay and sand-gravel. The marine beds roughly correspond to Ma 1 in Fig. 5.11 are exposed 500 m in height north of Osaka Bay, while beds of similar age have been detected by drilling the bay floor. The difference in altitude, about 1000 m, resulted from the Rokko Movement since 1 m. y.

The Ko-Biwako Group (ISHIDA and YOKOYAMA, 1969) of the similar age in the Biwa Lake district is not marine, but lacustrine and fluvial, and the Tokai Group (MAKINOUCHI, 1976) in the Nobi Plain to the east is lacustrine.

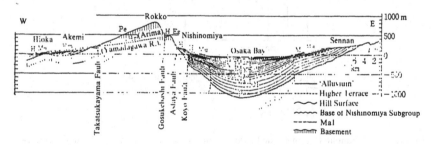

Fig. 5.10. Idealized cross-section of the Rokko Mountains—Osaka Bay—Sennan district (after HUZITA, 1979). Ol: Lower part of the Osaka Group, Ou: Upper part of the Osaka Group (Ma1 to Ma8: Marine clay beds, N: Nishinomiya Subgroup), Pe to M: Late Quaternary surfaces (Pe: Uplifted peneplain, E: Erosional surface. H: High terrace, Hm: Akami surface, Hr: High terrace of the Yamadagawa area, Hs: Shinobu-yama surface, M: Middle terrace, Mn: Nishiyagi surface, Mu: Uegahara surface).

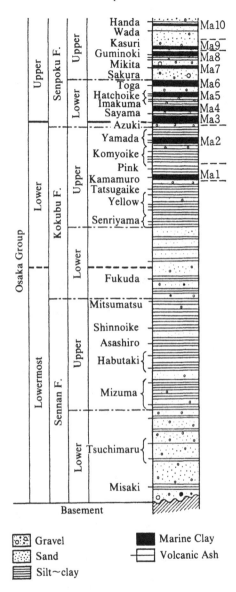

Fig. 5.11.   Columnar section of the Osaka Group in the Sempoku and Sennan districts (after ITIHARA *et al.*, 1975). The central column shows the names and horizons of the principal tephra beds. Ma1 to Ma10: Marine clay beds.

The Seto Group exposed in the area to the east of Nagoya City includes industrial porcelain clay deposits in its lower horizon.

5.8 Nankai-Tokai Province (Pacific Coast Areas)

The rocks of the Shimanto belt are overlain unconformably by Lower to Middle Miocene strata in several areas such as southern Kii Peninsula and the Shizuoka district. Granitic and rhyolitic rocks (13–15 Ma) were intruded into and extruded over the Miocene strata in such areas as the Sobosan, Okueyama and Osuzuyama districts in Kyushu, the Ishizuchi district of Shikoku, and the Omine and Kumano districts in the Kii Peninsula. These Miocene felsic rocks and other igneous rocks, are in general relatively rich in K_2O, and are similar to those of the Setouchi province.

a) Kii Peninsula (Fig. 5.12)

In the southern part of the peninsula the early Miocene (?) Muro Group are exposed, and consists mainly of turbidites, sandstone and mudstone. Rounded pebbles of orthoquartzite have been found in conglomerates of the upper horizons, and paleo-current studies suggest that materials were supplied from the south. It is inferred that a continental terrane would have existed to the south of the sedimentary basin of the Muro Group (HARADA and TOKUOKA, 1974). The Muro Group is unconformably overlain by the

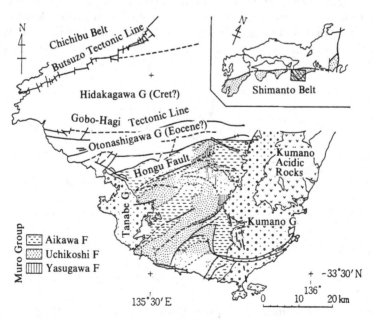

Fig. 5.12. Kii peninsula G: Group, F: Formation (after SUZUKI et al., 1979).

Tanabe Group in the west and the Kumano Group in the east, (TANAI and
MIZUNO, 1954) both groups are of Middle Miocene age.

b)   Shizuoka district (Fig. 5.13)

The Shimanto belt occurs in the outermost part of the Shizuoka district,
it is unconformably overlain by the early Miocene Oigawa, Kurami and
Saigo Groups. They consist of sandstone and mudstone deposited in a fully
marine environment and are associated with basic igneous rocks such as the
Takakusayama alkali basalt (TIBA, 1966) of the Oigawa Group.

The three groups are successively overlain by the late Miocene Sagara
and the Pliocene Kakegawa Groups. These strata are composed of sandstone
and mudstone with abundant shallow-marine fossils in some horizons.

The generalized correlation table of the Neogene strata of the Kanto,
southern Fossa Magna and Nankai-Tokai provinces is given in Table 5.5.

5.9   Kyushu

In northern Kyushu Neogene rocks are represented by coal-bearing
Miocene strata and alkali basalts of Pliocene age. In central Kyushu there
are no marine sediments, but volcanics and intrusive rocks of Miocene age
occur, and in southern Kyushu early Miocene and Pliocene sediments, and
Miocene to Quaternary volcanic rocks are widely exposed (Fig. 5.14).

The coal-bearing Lower Miocene strata, which lie conformably onto
Paleogene rocks, are exposed in a few areas of northern Kyushu, for
example, in the Sasebo-Karatsu coal field. They mainly consist of sandstone,
sometimes containing glauconite, and shale, and they carry abundant

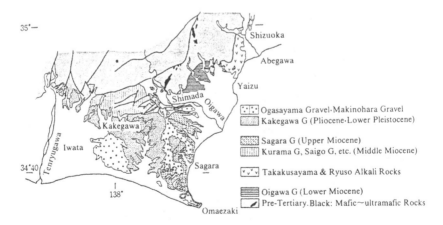

Fig. 5.13.   Shizuoka district. G: Group (based on the 1:200,000 Geological Maps "Shizuoka",
"Omaezaki" and "Toyohashi" issued by Geological Survey of Japan).

Table 5.5. Correlation of the principal Neogene strata of the Nansei-shoto, Nankai, southern Fossa Magna and Kanto areas. G: Group.

Age (Ma)		Plank. Foram. Z.	Nansei-Shoto	Nankai				Southern Fossa Magna			Kanto			
				Miyazaki	Shikoku	Kii	Shizuoka	Fujigawa	Tanzawa	Izu	Miura	Boso	Chichibu	Tomioka
0	Quat.	N23												
2		N22					Soga*				Urago*	Kazusa G*		
	Pliocene	N21	Shimajiri G* (Okinawa)	Miyazaki G*	Tonohama*		Kakegawa G*	Fujigawa G	Aikawa G / Nishikatsura G*	Ashigara G	Ikego*	Anno*		
5		N19 / N18				Sagara G*				Shirahama G	Zushi*	Kiyosumi*		
	Miocene Late	N17						Nishiyatsushiro G*						
10		N16										Amatsu*		
		N15											Kamiyokoze	Tomioka G*
		N14											Chichibu-machi G*	
15	Miocene Middle	N10 / N9							Tanzawa G					
		N8	Yaeyama G		Iorigawa	Misaki G / Kumano G	Saigo*				Hayama G	Kinone* / Akahira*		
	Miocene Early	N7					Kurami*					Hota G		
20		N6		Nichinan G		Muro G	Oigawa G			Yugashima G				

*Tsuchi (1979) —·— Ikebe (1978)

shallow marine shells and planktonic foraminiferas. Younger sediments of the Sasebo and Nojima Groups cover the Lower Miocene, particularly in the Sasebo area. The Sasebo Group consists of shallow-marine~brackish~ fresh-water sediments frequently intercalated with coal seams and tuff, and also contain abundant shells, foraminifera and fossil plants (NAGAHAMA, 1965). The Nojima Group is essentially terrestrial and rich in pyroclastic materials. Vast quantities of alkali basalts were subsequently erupted on the Miocene beds during Pliocene time (KURASAWA, 1967).

Miocene volcanic rocks similar to those of the Setouchi province are associated with the terrestrial sediments; the volcanic rocks of the Sekinan Group in the Beppu area are an example.

A part of the Nichinan Group of the Shimanto belt in southeastern Kyushu is considered to be of Lower Miocene age and consists of thick (over 1000 m) piles of marine sandstone and shale. It is unconformably covered by the similarly thick (2000–4000 m) Pliocene Miyazaki Group (SHUTO, 1952) consisting of intercalated marine sandstone and siltstone. On the other hand, the Miocene and younger rocks of southwestern Kyushu are mostly volcanic and not associated with marine sediment.

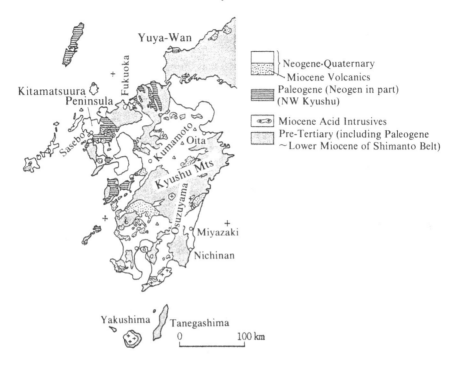

Fig. 5.14.   Kyushu (based on the 1:200,000 Geological Map of Japan).

5.10   Nansei-Shoto (Ryukyu archipelago)

A chain of small islands called the Ryukyu archipelago extends for about 1200 km from Kyushu to Taiwan. The chain is divided into three parts by the Tokara Straight and the Miyako Deep, into the northeastern, central and southwestern parts (Fig. 5.15) (KONISHI, 1965).

Miocene strata of the archipelago are mostly exposed on the Tanegashima and Yakushima islands of the northeastern part and on the Iriomote and Yonaguni islands of the southwestern part (KIZAKI and OSHIRO, 1977). The strata of the latter islands consist of the Yaeyama Group of shallow-sea~brackish sandstone, siltstone and conglomerate.

Late Miocene to Pliocene strata consisting of marine siltstone, sandstone and dolostone are termed the Shimajiri group and are distributed in the Kikai, Kume and Miyako islands and also in the southern part of Okinawa main island.

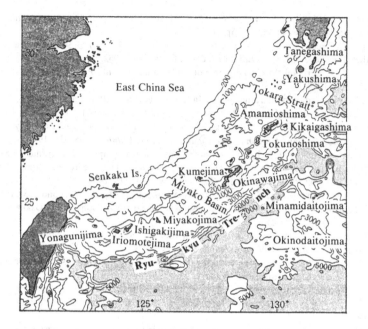

Fig. 5.15.   Nansei-shoto.

REFERENCES

CHINZEI, K., Younger Tertiary geology of the mabechi river valley, northeast Hokkaido, Japan, *J. Fac. Sci., Univ. Tokyo, II*, **16**, 161–208, 1966.

CHINZEI, K., Neogene molluscan faunas in the japanese Islands: an ecologic and zoogeographic synthesis, *The Veliger*, **21**, 155–170, 1978.

COLLABORATIVE RESEARCH GROUP FOR THE SHIDARA BASIN, The Neogene distributed in the south district of Mt. Horaiji, Aichi Prefecture, central Japan, *Mem. Geol. Soc. Japan*, **16**, 63–75, 1979.

FUJII, K. and SOGABE, M. Tectonic movement occurred in Hokkaido during the late Miocene and Pliocene time, *Bull. Geol. Surv. Japan*, **29**, 631–644, 1978.

FUJII, S., SOMA, T., OTSUKA, T., KONDO, Y., OGAWA, K., SAKAMOTO, S., and ASANO, H., Explanatory Text of the Geological Map of Toyama Prefecture, Toyama Prefectural Office, 1–12, 1970.

FUJITA, T., Volcanostratigraphy of the Misasa group in the vicinity of Ningyo-toge, *J. Geol. Soc. Japan*, **78**, 13–28, 1972.

FUJITA, T., MATSUMOTO, Y., SHIMAZU, M., and WADATSUMI, K., Volcanostratigraphy of the Neogene formations in Southwest Japan and the Fossa Magana Region, *Cenozoic Geology of Japan (Professor Nobuo Ikebe Volume)*, 121–133, 1978.

HARATA, T. and TOKUOKA, T., Kuroshio Paleoland, *Kagaku (Science)*, **44**, 495–502, 1974.

HUZIOKA, K., OZAWA, A., and IKEBE, V., *Geology of the Ugo-Wada district*, Geological Survey of Japan, 65p., 1976.

HUZIOKA, K., OZAWA, A., TAKAYASU, T., and IKEBE, Y., *Geology of the Akita district*, Geological Survey of Japan, 75p., 1977.

HUZITA, K., Tectonic development of the Median Zone (Setouchi) of Southwest Japan, since the Miocene, *J. Geosci., Osaka City Univ.*, **6**, 103–144, 1962.

HUZITA, K., Criteria on the working time of active faults in Kinki Province, *Earth Monthly*, **1**, 583–591, 1979.

IKEBE, N., Bio- and chronostratigraphy of Japanese Neogene, with remarks on paleogeography, *Cenozoic Geology of Japan, Professor Nobuo Ikebe Memorial Volume*, 13–34, 1978.

IKEBE, Y., KTAHIRA, T., and MIYAZAKI, H., Some problems on petroleum geology in Japan, *Cenozoic Geology of Japan (Professor Nobuo Ikebe Volume)*, 205–216, 1978.

ISHIDA, S. and YOKOYAMA, T., Tephrochronology, paleogeography and tectonic development of Plio-Pleistocene in kinki and Tokai district, Japan—The research of younger Cenozoic strate in Kinki Province, Part 10, *Quatern. Res.*, **8**, 31–43, 1969.

ITIHARA, M., YOSHIKAWA, S., INOUE, K., HAYASHI, T., TATEISHI, M., and NAKAJIMA, K., Stratigraphy of the Plio-Pleistocene Osaka standard stratigraphy of the Osaka Group, *J. Geosci., Osaka City Univ.*, **19**, 1–29, 1975.

ITOIGAWA, J. and SHIBATA, H., Paleoenvironmental change and correlation based on molluscan assemblages, *Mem. Geol. Soc. Japan*, **8**, 125–136, 1973.

KAKIMI, T., KINUGASA, Y., and KIMURA, M., Neotectonic Map: Tokyo 1:500,000, Tectonic Map Series 2, Geological Survey of Japan, 1973.

KASENO, Y. (ed.), Environmental Geology of Ishikawa Prefecture (The explanatory text of the geological map on 1:100,000), *Natural Environments of Ishikawa Prefecture*, **1**, 1–128, 1977.

KATAHIRA, T., Stratigraphy of the Neogene Tertiary in the central and northern parts of Niigata Prefecture—Petroleum geology of the Neogene Tertiary in the chuetsu and the Kaetsu regions, Niigata, Japan, *J. Japan. Assoc. Petrol. Tech.*, **39**, 167–178 and 337–356, 1974.

KAWABE, T., FUKUDA, T., KAWABATA, N., and MAEDA, S., The late Cenozoic kurotaki Formation distributed from the Minato River-area to the Obitsu River-area, central Boso Peninsula, *J. Geogr.*, **88**, 281–295, 1979.

KITAMURA, N., TAKAYANAGI, Y., MASUDA, K., HAYASAKA, S., MITSUI, S., SUGAWARA, K., and TAKAHASHI, K., On some problems concerning the Tertiary strata of the Izu peninsula, Japan, *Contrib. Inst. Geol. Paleontol., Tohoku Univ.*, **75**, 19–31, 1969.

KITAZATO, H., Geology and geochronology of the Younger Cenozoic of Oga Peninsula, *Contrib. Inst. Geol. Paleontol., Tohoku Univ.*, **75**, 17–49, 1975.

KIZAKI, K. and OSHIRO, I., Paleogeography of ;the Ryukyu islands, *Marine Sciences/Monthly*, **9**, 542–549, 1977.

KONISHI, K., Geotectonic framework of the Ryukyu islands (Nansei-shoto), *J. Geol. Soc. Japan*, **71**, 437–457, 1965.

KOSAKA, T. and TSUNODA, F., Geology of the Koma mass if in the western part of Yamanashi Prefecture, *J. Geol. Soc. Japan*, **75**, 127–140, 1969.

KURASAWA, H., Petrology and chemistry of the Amagi volcanic rocks, Izu Peninsula, Japan, *Chikyu Kagaku (Earth Sciences)*, **44**, 1–18, 1959.

KURASAWA, H., Petrology of the Kita-Matsuura basalts in northwest Kyushu, Southwest Japan, *Rept. Geol. Surv. Japan*, **217**, 1–108, 1967.

MAKINOUCHI, T., Geologic structures of the southern part of the Chita peninsula and the tectonic ;movements around the Ise bay, central Japan, *J. Geol. Soc. Japan*, **82**, 311–325, 1976.

MATSUDA, T., The Miocene stratigraphy of the Fuji river valley, *J. Geol. Soc. Japan*, **67**, 79–96, 1961.

MATSUDA, T., Crustal deformation and igneous activity in the South Fossa Megna, Japan, *Geophys. Monogr. Amer. Geophys. Union*, 6, 140-150, 1962.

MATSUNO, K., *Geology of the Wakayanagi district*, Geological Survey of Japan, 24p., 1967.

MATSUNO, K. and YAMAGUCHI, S., 1:50,000 Geological Map "Shizunai", Geol. Surv. Hokkaido., 1958.

MIKAMI, K., Geological and petrological studies on the Tanzawa mountainland, *Sci. Rept., Yokohama Natn. Univ., Part I, Ser. II*, 8, 57-110, 1961.

MITANI, K., Changing of the Tertiary sedimentary basins in the western flank of the axial belt of Hokkaido—Significance of the Sunagawa lowland-the Umaoi hilly area belt, *Assoc. Geol. Collabor. Japan Monogr.*, 21, 127-137, 1978.

MITSUNASHI, T., The sedimentary units in relation to the sedimentary process of the Neogene formations in the south Kanto, central Japan, *Professor Kazuo Huzioka Volume*, 249-260, 1977.

MIYASAKA, S. and KIKUCHI, K., The Neogene Tertiary upheaval of the Hidaka metamorphic belt, *Assoc. Geol. Collabor. Japan Monogr.*, 21, 139-153, 1978.

MIZUNO, A., SUMI, Y., and YAMAGUCHI, S., Miocene stratigraphy of the Kushiro coal-field, eastern Hokkaido, with special reference to the stratigraphic problem concerning the so-called Chokubetsu Formation, *Bull. Geol. Surv. Japan*, 20, 633-649, 1969.

NAGAHAMA, H., Diagonal bodding and accumulation of Tertiary sediments in northwestern Kyushu, Japan, *Geol. Surv. Japan Rept.*, 211, 66p., 1965.

NAKADA, S. and TAKAHASHI, M., Regional variation in chemistry of the Miocene intremediate to felsic magmas in the Outer Zone and the Setouchi province of Southwest Japan, *J. Geol. Soc. Japan*, 85, 571-582, 1979.

NAKAGAWA, H., ISHIDA, T., OIKE, S., ONODERA, S., TAKEUCHI, S., NANASAKI, O., MATSUYAMA, T., and TOGA, T., Quaternary crustal movements in the Shinjo Basin, northeastern Honshu, Japan, *Contrib. Inst. Geol. & Paleontol., Tohoku Univ.*, 71, 13-29, 1971.

NIITSUMA, N., Magnetic stratigraphy in the Boso peninsula, *J. Geol. Soc. Japan*, 82, 163-181, 1976.

OZAWA, A., FUNAYAMA, Y., and KITAMURA, N., *Geology of the Kawashiri district*, Geological Survey of Japan, 40p., 1971.

OTA, Y. and SUZUKI, I., Notes on active folding in the lower reaches of the Shinano River, central Japan, *Geogr. Rev. Japan*, 52, 592-601, 1979.

OTSUKI, K., Geology of ;the Tanakura Shear Zone and ajacent area, *Contrib. Inst. Geol. Paleontol., Tohoku Univ.*, 76, 1-71, 1975.

SAKAMOTO, T., Cenozoic strata and structural development in the southern half of the Toyama basin, central Japan, *Geol. Surv. Japan. Rept.*, 213, 27p., 1966.

SAN'IN GREEN TUFF RESEARCH GROUP, On the collapse basins of the lower and the middle Miocene series at the south of Matsue City, southwest Japan, *Mem. Geol. Soc. Japan*, 16, 33-53, 1979.

SEKI, Y., OKI, Y., MATSUDA, T., MIKAMI, K., and OKUMURA, K., Metamorphism in the Tanzawa mountains, central Japan, *J. Japan. Assoc. Mineral. Petrol. Econ. Geol.*, 61, 1-21 and 50-75, 1969.

SHUTO, T., Stratigraphical study of the Miyazaki Group, *Sci. Rept., Fac. Sci., Kyushu Univ., Geology*, 4, 1-40, 1952.

SUGIMURA, A., Uniform rates and duration period of Quaternary Earth movements in Japan, *J. Geosci., Osaka City Univ.*, 10, 25-35, 1967.

SUZUKI, H., HARATA, T., ISHIGAMI, T., KUMON, F., NAKAYA, S., SAKAMOTO, T., TATEISHI, M., TOKUOKA, T., and INOUCHI, Y., *Geology of the Kurisugawa district*, Geological Survey of Japan. 54p., 1979.

SUZUKI, K., MANABE, K., and YOSHIDA, T., The Late Cenozoic stratigraphy and geologic development of the Aizu basin Fukushima Prefecture, Japan, *Mem. Geol. Soc. Japan*, **14**, 17–44, 1977.

TAGUCHI, K., Basin architecture and its relation to the petroleum source rocks development in the region borering Akita an Yamagata Prefectures and the adjoining areas, with the special reference to the depositional environment of petroleum source rocks in Japan, *Sci. Rept., Tohoku Univ., Ser. III*, 7, 293–342, 1962.

TAI, Y., Miocene microbiostratigraphy of west Honshu, Japan, *J. Sci., Hiroshima Univ., Ser. C*, **2**, 265–395, 1959.

TAI, Y., On the "Shinji Folded Zone", *Mem. Geol. Soc. Japan*, **9**, 137–146, 1973.

TAI, Y., Some problems on the peneplanetaion of the Chugoku mountains base on the Miocene paleogeography, *Japan, J. Geol. Geogr.*, **84**, 133–139, 1975.

TANAI, T. and MIZUNO, A., Geological structure in the vicinity of the Kumano coal field in southeastern Kii Peninsula, *J. Geol. Soc. Japan*, **60**, 28–39, 1954.

TIBA, T., Petrology of the alkaline rocks of the Takakusayama district, Japan, *Sci. Rept., Tohoku Univ., Ser. III*, **9**, 541–610, 1966.

Tokachi Research Group, Tokachi Plain, *Assoc. Geol. Collabor. Japan Monogr*, **22**, 433p., 1978.

TOYONO COLLABORATIVE RESEARCH GROUP, Quaternary geology and geological history in the western marginal area of the Nagano basin, *Mem. Geol. Soc. Japan*, **14**, 79–92, 1977.

TSUCHI, R. (ed.), *Fundamental data on Japanese Neogene bio- and chrono-stratigraphy,.* Shizuoka University, 156p., 1979.

TSUNEISHI, Y., Geological and experimental studies on mechanism of block faulting, *Bull. Earthq. Res. Inst.*, **53**, 173–242, 1978.

UEMURA, F., SAKAMOTO, T., and YMADA, N., *Geology of the Wakasa district*, Geological Survey of Japan, 91p., 1979.

UEMURA, T., Tectonic development of the Miocene sedimentary basins of east Mino, central Japan, *J. Earth Sci., Nagoya Univ.*, **9**, 394–417, 1961.

UI, H., Geologic structure of the Miocene Tomikusa Basin, Anan-cho, Shimoina-gun, Nagano Prefecture, Japan, *J. Geol. Soc. Japan*, **75**, 131–142, 1970.

YAMAGISHI, H., Collapse structure in sedimentary basin around Nukabira, central Hokkaido, Japan, *Rep. Geol. Surv. Hokkaido*, **48**, 21–31. 1966.

YAMAGUCHI, S., Neogene Tertiary of "Toyokoro-Kitami Belt"—Neogene Tertiary stratigraphy in the Tsubetsu-Honbetsu district, *Bull. Geol. Surv. Japan*, **17**, 673–681, 1966.

CHAPTER 6

QUATERNARY SEDIMENTS OF LOWLAND PLAINS AND TERRACED UPLANDS

Quaternary marine soft sediments are distributed mostly in areas of Neogene sedimentary basin development, where the former successively overlies the latter. These areas form geomorphic basins and lowland plains, and they are still subsiding to-day. The Quaternary sediments exhibit a change of facies from neritic-littoral to estuarine-fluvial or from lacustrine to fluvial, reflecting the depositional development of the regions. Furthermore, the effects of glacial eustacy are observed in the sediments younger than Middle Pleistocene, particularly in the coastal plain areas.

Japanese Quaternary sediments have the following distinctive characteristics. First, they are intercalated with a number of bands of volcanic materials, which play an important role in the correlation of the geomorphic surfaces or of stratigraphic horizons and also in the paleomagnetic stratigraphy. Second, coarse clastics predominate in the non-volcanic sediments due to the severe crustal movements of this period. Thick piles of fan conglomerates in intermontane basins, and coarse sediments in the estuaries are examples. Third, reef limestones are widely developed, particularly in the Nansei-shoto archipelago. Finally, fossils of the Japanese Quaternary are mostly terrestrial being comprised of fossil plants and large mammals, particularly elephants.

6.1 Terrace Deposits and Recent Formations

The middle to Late Pleistocene (500,000 to 20,000 years B.P.) terrace deposits and the lowland plain deposits of late Pleistocene to Holocene (younger than 20,000 years), are the principal concerns of this chapter.

6.1.1 Terrace deposits

Representative terrace surfaces are tabulated in Table 6.1. In the table the surfaces are designated with symbols such as Tc and S, and the deposits

Table 6.1   Classification of terrace surfaces of South Kanto region.

Terrace Surface	Symbols	Age[3] (10^4 y·ago)	Deposits
Tachikawa S[1]	Tc $\begin{cases} Tc_3 \\ \wr \\ Tc_1 \end{cases}$	1.5 $\wr$ 3	Tachikawa Gravel Bed
Misaki S	M	6	Misaki F
Obaradai S	O	8	Obaradai F
Shimosue-yoshi F	S	12~13	Shimosueyoshi F
Tama S	T $\begin{cases} T\text{-}a^{2)} \\ \wr \\ T\text{-}e \end{cases}$	15 $\wr$ 40	'Byobugaura F'

1) Fluvial plains.
2) Subdivision of T surfaces based on Machida et al. (1974).
3) ^{14}C method for Tc surfaces. Others based on fission-track method or inferred from sedimentation rates.

Table 6.2.   Correlation of Middle to Upper Quaternary of Japan. G: Group, F: Formation.

Age			Geomor. Surface	Nansei-shoto[1]	Southwest Japan		
					Miyazaki Plain[2]	Nobi Plain[3]	Kanazawa[4] ·Noto
$\times 10^4$	Holocene		Alluvial Plain	Emer. Reef	Marine Alluvium	Nanyo F	Sand Dune Deposits
1						Nobi F	Lagoonal Deposits
2	Pleistocene	Late	—Tc$_2$—		Fluvial Terrace Gravel Bed	Diichi GB	
3			—Tc$_1$—			Kamaki GB	
6			—M—	Araki Limestone	Kawamina-mibaru F	Upper	Fluvial Terrace Gravel Bed
8			—O—	Ryukyu Limestone / Ryukyu Group	----↑----	Atsuta F	
13			—S—		Nyutabaru F		
					Sanzaibaru F	Lower	Hiradoko F
15			—T—				
		Middle	Surfaces	Naha Limestone	Chausubaru F	Daini Gravel Bed	Higher Terrace Deposits
						Ama F	Utatsuyama F
						Yagoto F	
		Early		Shimajiri Group		Karayama F	Okuwa F

1) Konishi et al. (1974), 2) Hoshino (1971), 3) Nobi Plain Quaternary Research Group (1977), 4) Hokuriku Quaternary Research Group (1969), 5) Tusuchi (1960), 6) Toyono

with names such as the Misaki Formation and the Shimosueyoshi Formation. The most widely developed surface is the S surface, and consequently, the Shimosueyoshi marine Formation and correlatives are also the most widely spread of the terrace deposits, having been formed by the Shimosueyoshi Transgression which resulted from the sea-level rise of the last interglacial stage. Deposits of the Tc surface are generally fluviatile in the southern Kanto Plain.

Terrace deposits are generally barren of fossils, therefore correlations are mostly made on the basis of the tephra deposits, particularly those which widely cover the terrace deposits.

6.1.2 Recent formations

The Recent formations form the lowland plain, which can be traced without geomorphological discontinuity to the present-day base datum plane (e.g. sea-level).

Representatives of the Recent formations are the marine sediments underlying the coastal plains, which have been deposited by the Yurakucho Transgression which commenced about 20,000 years ago (KAIZUKA et al., 1977).

Table 6.2. (continued.)

S Foss. Mag. & Kanto		Northeast Japan			Hokkaido
Tokyo·Yokohama	Shizuoka Area[5]	Oga Peninsula	N Foss. Mag.[6]	Sendai Area[7]	Kushiro Plain[8]
Yurakucho F (Nanagochi F)	Tomoegawa Alluvium	Hachirogata Alluvium	Nojiriko F	Marine Alluvium	Soft Ground
Tachikawa GB Ekoda F				Nakamachi Ter. GB	Younger LTG
				Kamimachi Terrace Gravel Bed	Harutori F
Musashino Gravel Bed	Kuniyoshida Gravels / Oshika Gravels	Terrace Gravel Bed	Terrace Gravel Bed		Older Lower Terrace Gravel
Shimosueyoshi F	Kusanagi Mud	Katanishi F	Minamigo F	Dainohara Gravel Bed	Otanoshike F
Sagami G	Kunozan Gavels	Shibikawa F	Toyono F	Aobayama F	Kushiro G
Kazusa G	Nekoya Mud	Wakimoto F	Shigarami F		

COLLABORATIVE RESEARCH GROUP (1977); NOJIRI-KO EXCAVATION RESEARCH GROUP (1970), 7) NAKAGAWA et al. (1960), 8) MINATO et al. (1972).

6.2   Hokkaido

In Hokkaido thick Quaternary sediments are distributed in the Ishikari Lowland, the Tokachi and Kushiro Plains and on the Konsen terraced upland. In addition, mountain gravels (Lower Pleistocene) are observed in intermontane basins, and marine terraces (Middle to Upper Pleistocene) have been developed almost completely around the island.

6.2.1   Ishikari Lowland

Terraces comprising a few steps have been developed in the lowland, although their correlation to the main representatives (Table 6.1) has not yet been worked out. The marine Recent formations reach 140 m in thickness; the lower part consists of clay, sand and gravel, and the upper part of peat and dune sand.

6.2.2   Tokachi Plain

Fan gravels younger than Middle Pleistocene widely cover the Lower Pleistocene Osarushinai Formation, forming a few terraces. Recently, bones of *Elephus namadicus naummanni* were discovered in peaty lacustrine deposits, probably of the Shimosueyoshi time, in the southern part of the Toyokoro Hills district. On the other hand, a molar of *Mammontheus primigenius* was obtained from a gravel bed of the Ogoshi terrace, which was covered by the First Shikotsu Pumice Fall Deposit (about 32,000 ^{14}C age) and this is correlated with the M surface.

6.2.3   Konsen terraced upland

Lower to Middle Pleistocene sediments (>500 m thick), were deposited in this district. They are termed the Kushiro Group and consist of intercalated transgressive and regressive sediments. Its upper boundary is the Nemuro surface, 140–100 m in altitude, and this is dissected by the Konsen upland plain, 30–70 m. Furthermore, the Kushiro Group is partly covered by the Late Pleistocene Otanoshike Formation rich in marine shell remains, whose depositional surface (Kushiro surface) is correlated with the S surface.

Recent formations, about 100 m in thickness, underlie the surface of the coastal plain of the Kushiro district. The lower part consists of sand, gravel and clay, the middle part of marine clay containing shells and foraminifers, and the upper part consists of dune sand, peat and gravel (MINATO *et al.*, 1972).

6.3 Tohoku Province and Contiguous Regions

6.3.1 Sea of Japan coastal areas

Thick sediments, partly of marine facies, of Middle to Upper Pleistocene age are particularly developed in the northern to central Akita district and in the Niigata Plain. In the former district the Middle Pleistocene Shibikawa Formation covers the lowermost Pleistocene Wakimoto Formation. The former is of tuffaceous sand, but includes some gravel, lignite, marine shells and further elephant bones. It is overlain by the Katanishi Formation, whose lower part carries marine shells. Its depositional surface corresponds to the S surface. A number of coastal terraces lower than 300 m in elevation are developed in the Tsugaru Peninsula, Shiragami Hills and Oga Peninsula. Their surfaces are upwarped, suggesting local crustal movement is still active (OTA, 1975).

In the marginal part of the Niigata Plain, the Middle Pleistocene Yashirota Formation unconformably covers the Uonuma Group of Upper Pliocene to Lower Pleistocene age. The Yashirota Formation consists of thick piles of brackish to fresh-water sand, gently folded together with the underlying strata. The river terraces in this district also underwent the folding movement. On the other hand, Recent formations underlying a substantial part of the plain comprise the Shirone Formation, which reaches 140 m in maximum thickness. Its lower and upper parts are composed of brackish~fresh-water sand-mud and sand-gravel, while the middle part consists of marine clay.

6.3.2 Intermontane basins

A number of fault-angle basins are located between the Dewa Hills and the Backbone Range. They are, from north to south, the Odate, Takanosu, Yokote, Shinjo, Yamagata, Yonezawa and Aizu Basins. Thick fluvial gravel beds resulting from fans and flood plain deposits occur in their central parts. Tuffs, pyroclastic flow deposits and lignite are occasionally intercalated with the gravels, the lignites yielding abundant plant remains. Detailed studies have been made, particularly on the Quaternary deposits of the Aizu Basin (SUZUKI et al., 1977).

Similar intermontane basins are also developed between the axial range and the Kitakami and Abukuma Plateaus to the east. They are, from north to south, the Kitakami, Fukushima and Koriyama Basins (Fig. 6.1). The basins are underlain by a number of fan gravel beds, which are covered by tuff. The gravel beds occasionally yield plant fossils and large mammals. Particularly noticed is the Hanaizumi Formation (20,000–35,000 years old), whose depositional top corresponds to the Tc_2 surface. It contains abundant

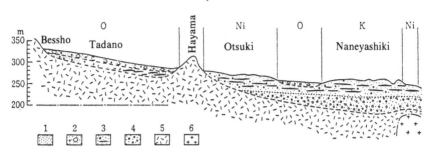

Fig. 6.1.   Topographic and geologic cross section of the Koriyama Basin (SUZUKI *et al.*, 1967).
L: Oharuda terrace surface, S: Shin-yashiki terrace surface, O: Otsuki surface, N: Nishinouchi
surface, K: Koriyama surface, 1: Lower terrace and Shin-yashiki terrace deposits, 2: Otsuki

fossils of cold-climate plants, steppe mammals such as *Elephas naummanni*, *Megaloceros* and bisons, and *Alces*, the last being of the mammoth fauna.

Furthermore, in the Nagano Basin of the northern Fossa Magna region the marine Lower Pleistocene is covered by the lacustrine Middle Pleistocene Toyono Formation (TOYONO FORMATION RESEARCH GROUP, 1977). They are folded together and unconformably overlain by the lacustrine Minamigo Formation. All of the formations are further covered by fan and terrace gravel beds. The Nojiriko Formation (16,000–40,000 years old) on the Nojiriko lake floor yields a number of bones such as those of *Elephas naummanni* and *Megaloceros yebei* in association with paleoliths (NOJIRIKO SURVEY GROUP, 1975).

6.3.3   Pacific coast areas

Wide coastal terraces of Middle to Late Pleistocene age are developed in the eastern Aomori district. They are underlain by Pliocene to Lower Pleistocene marine strata and overlain by tuff. The terraces are classified into the higher Tengutai plane (=T surface) and the lower Takadate plane (=S surface).

In the Sendai District, terraces of several steps dissect the marine Pliocene, the highest of which is the Aobayama terrace (=T). Along the Pacific coast, to the east of the Abukuma Plateau, many terraces are also developed of which the widest terrace corresponds to the S surface. Paleomagnetic study of terrace silts indicates that the time of the S surface corresponds to the Blake event (ca. 110,000 y.B.P.).

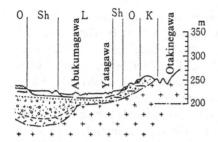

Formation, 3: Upper part of the Koriyama Formation, 4: Lower part of the Koriyama Formation, 5: Pre-Upper Quaternary, 6: Granites.

6.4   Kanto Plain and Southern Fossa Magna Region

6.4.1   Kanto Plain

The Middle Pleistocene Sagami Group unconformably overlies the Kazusa Group in the northern Miura Peninsula. In its lowermost part is the muddy Naganuma Formation, which has been well known for yielding abundant shallow marine shell fossils. The Naganuma Formation is covered by the Byobugaura Formation, consisting of several sediment cycles commencing with a marine facies and ending with an aeolian tuff facies (Tama Loam) (MACHIDA et al., 1974). The Sagami Group becomes thicker westwards to the Oiso hills, suggesting the existence of a sedimentary basin there. The Oshinuma sand and gravel beds of the middle part of this group covers the abrasion platform of the Kazusa Group. The depositional plane T-c is about 0.3 Ma, as judged by fission track age of pumice covering the bed.

In the northern part of the Boso Peninsula the Middle to Upper Pleistocene Shimosa Group reaches 400 m maximum thickness and unconformably covers the Kazusa Group. The sediments are mostly shallow marine sand with occasional clay layers. A number of shell fossil localities are known. The temperature of sea water during time of deposition has been inferred as 10–20°C by means of the oxygen isotope method (MASUDA and TAIRA, 1974). The column ranging from the lowermost Jizodo Formation to the upper Kamiiwahashi Formation is considered to be correlated with the Sagami Group described above.

Marine Upper Pleistocene strata are most widely developed in the Kanto Plain as compared with other Japanese districts. The widest plain is the S surface, whose altitude is less than 10 m in the central part and higher than 100 m in the peripheral part of the Kanto Plain, suggesting the basinal development of the district. The Shimosueyoshi Formation of the S surface consists of lower muddy beds and upper sand-gravel beds. The former fills the valley dissected into the basement and the latter covers the abrasion platform 20–40 m in altitude (Fig. 6.2). The correlative upper Tokyo Formation and Kioroshi Formation are sediments of the Paleo-Tokyo Bay. They contain shells of the bay environment and bones of *Elephas naummanni*. The paleo-temperature inferred from fossil evidence was cooler in the early and late periods of the transgression-regression cycle and warmer in the middle. Therefore, a cycle of climatic and sea-level changes is suggested, and this is correlated with the Riss/Würm or Sangamon inter-glacial stage.

Fluvial terraces corresponding to the O and M surfaces are developed in the peripheral parts of the Kanto Plain. The Tachikawa Terraces, on the other hand, are classified into Tc_1, Tc_2 and Tc_3, according to their heights and the horizons of the Tachikawa Loam (ash) member covering them. The terrace surface Tc_1 composed of the Tc_1 gravel bed is covered by the whole Tachikawa Loam member. It was formed about 30,000 years ago, when the Egota conifer bed yielding subalpine needle-leaves were deposited. The planes Tc_2 an Tc_3 are respectively about 20,000 and 15,000 years old. Their gravel beds lie underneath the alluvial lowland plains in the lower reaches areas and are further connected with the fluviatile gravel bed of the Paleo-Tokyo Bay within the Tokyo Bay district.

The Yurakucho Formation is divided into two parts, each comprising a

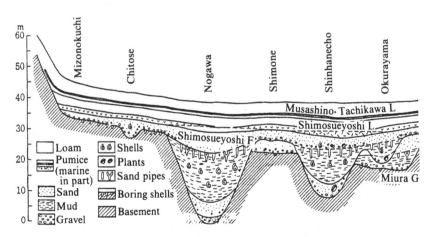

Fig. 6.2.   Idealized cross-section of the Shimosueyoshi terraced upland. G: Group, F: Formation, L: Loam.

sedimentary cycle. A sand bed is intercalated between them (Fig. 6.3). The lower part represents the early stage of the Yurakucho Transgression, while the upper part represents the maximum transgression to regression, and the middle sand bed indicates the stillstand stage, 11,000 years ago. The upper part consists of soft clay containing marine shells. It extends widely further into the inner part of the coastal plain, suggesting a transgression, called the Jomon Transgression.

Holocene coastal terraces are well developed in the Oiso Hills, the Miura Peninsula and the southern Boso peninsula. The highest is the Numa Terrace of uplifted shore sediments, exemplified by the Numa coral bed. Its age is suggested to be 6,000–8,000 years. In the southern part of the Boso Peninsula four terraces are developed outside of the Numa terrace. The shore-line heights are respectively 16 m (3,600 years), 12 m (2,900 years), 5 m (270 years) and 1–2 m (1923: the time of the Great Kanto Earthquake). The altitudes of the coastal terraces was adjusted by each great earthquake.

6.4.2 Southern Fossa Magna

In the Iwabuchi Hills district, the Saginota gravel bed of bay mud and river gravel covers the Lower Pleistocene strata. It has yielded bones of *Stegodon orientalis*. The Kusanagi mud bed is correlated with the Shimosueyoshi Formation in the Udoyama area near Shizuoka City. It carries abundant bay shells and plant remains. Beneath the bed lies the Kunozan fan gravel, which has yielded bones of *Elephas naummanni*.

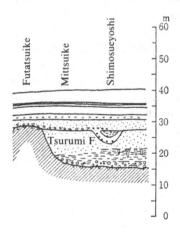

Fig. 6.2.   (continued.)

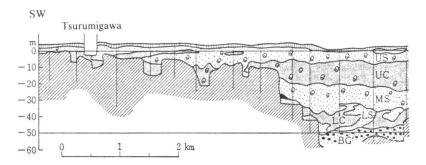

Fig. 6.3. Alluvial strata in the lower reach area of the Tamagawa river (after KAIZUKA *et al.*, 1977). US: Upper Sand, UC: Upper Clay, MS: Middle Sand, LC: Lower Clay, LS: Lower Sand, BG: Basal Conglomerate.

6.5 Southwest Japan

6.5.1 Pacific coast areas

Upper Pleistocene sediments are widely distributed in the Tokai district, mostly forming the S and O surfaces, while in the areas from the Kii Peninsula to Shikoku they are represented by only a thin veneer of gravel beds on the wave-cut terraces. In southern Kyushu, Middle to Upper Pleistocene strata are widely distributed again.

In the lower reaches of the rivers Oi and Tenryu, terraced uplands consisting of thick developments of fan sand and gravel similarly corresponds to the S and O surfaces. They are underlain by marine sediments correlated with the Shimosueyoshi Formation. The Furuya mud bed of the Makinohara upland and the Sahama mud bed of the Mikatagahara upland are examples. Sediments of open sea-shore form the coastal terraces in the area extending from the Atsumi to Shima Peninsulas. Coastal terraces, comprising three steps were also developed in southern Kii Peninsula and the areas around Tosa Bay, Shikoku. The elevated former shore lines of the Kii Peninsula become lower northwestwards. This tendency is stronger in higher than in lower terraces, and shows the positive correlation with the seismic displacement that occurred during the time of great earthquakes in the Nankai Trough region. It is suggested that the tilting may have been accumulated during each earthquake displacement (YOSHIKAWA *et al.*, 1964). The middle terrace is the widest, and its deposits are correlated with the Shimosueyoshi Formation. The lowest terrace is of Holocene age.

A few terraces lower than 270 m are developed in the Miyazaki Plain, southeastern Kyushu. Sediments of the higher Chausubaru plain consist of lower marine and upper fluviatile sand-gravel. The Sanzaibaru plain

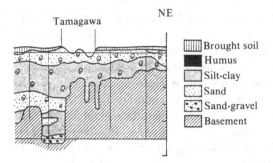

Fig. 6.3. (continued.)

corresponds to the S surface, is composed of marine sand-gravel. Further-more, the Nyutabaru gravel bed of the Nyutabaru plain, which is correlated with the O or M surfaces, is fluviatile. The high rate of upheaval in this district is suggested by the former shore line being generally higher and the occurrence of buried valleys at the base of the sediments (Fig. 6.4).

6.5.2 Setouchi

Structural basins such as the Osaka and Nobi Plain districts are the principal distribution areas of marine Quaternary in the Setouchi region.

The high, middle and low terrace groups of fan gravels are developed in the peripheries of the Osaka, Harima, Kyoto, Nara and Omi basinal areas. The upper part of the higher terrace deposits is formed by the Harima

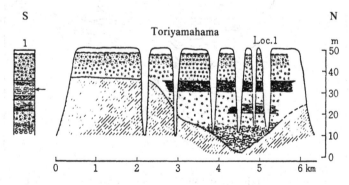

Fig. 6.4. Cross-section of the sea-cliff at Kawaminamibaru in the miyazaki plain (after HOSHINO, 1971). Loc. 1: Locality of the ^{14}C-dated sample. The columnar section at Loc. 1 is given to the left. The arrow indicates the collecting point.

Formation, and the lower part is formed by the Kiyotani Formation in the
Rokko area, both consisting of red weathered gravels (ITIHARA, 1960). The
Uemachi Formation of the middle terrace includes marine clay beds and
yields shells and bones of *Elephas naummanni*. The low terrace gravel bed,
10,000–30,000 years old, carries cold-climate plant remains. The marine
Recent formations beneath the Osaka basinal plain are called the Namba
Formation, and its basal part consists of peaty materials about 20,000 years
old. It shows the Yurakucho Transgression. Terraces of sand and gravel are
developed around Lake Biwa, but the 200 m drilling of the lake floor at the
center of the lake showed that all the recovered sediments were clay.

Terraces are developed mostly in the eastern part of the Nobi Plain
district, according to the tilting movement of this region. The oldest are the
non-marine Karayama and Yagoto Formations, and the latter seem to be
middle Pleistocene. Only the middle terraces such as the Atsuta and
Kagamihara terraces are marine. The upper part of the Atsuta Formation
includes the Third Ontake Pumice (Pm-III, 50,000~60,000 y.B.P.). The
intercalations of marine clay and sand-gravel younger than Middle Pleistocene
underlie the Nobi Plain, suggesting the repetition of transgression and
regression. Of these the Lower Atsuta Formation is the most widely spread
and indicates the Shimosueyoshi Transgression (QUATERNARY OF NOBI
PLAIN DISTRICT STUDY GROUP, 1977; Fig. 6.5). The First Nobi Gravel Bed
covering the Atsuta Formation seems to be connected with the lower terrace
gravel and is considered to be indicative of the Würm stage. The marine

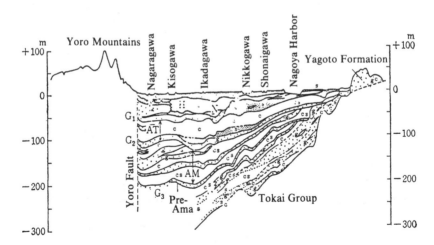

Fig. 6.5. Cross-section showing the sub-surface geologic structure of the Nobi plain. H: Nan-
yo Formation, N: Nobi Formation, G_1: The First Gravel Bed, AT: Atsuta Formation, G_2: The
Second Gravel Bed, AM: Amabe Formation, G_3: The Third Gravel Bed, c: Clay, s: Sand, g:
Gravel.

Recent deposits are classified into the upper Nobi and the lower Nanyo Formations. Peaty clay at the base of the former is 16,000– 18,000 ^{14}C years old and indicates the beginning of the Yurakucho Transgression. Shallow buried valleys formed in the upper part of the Nanyo Formation show that the sea-level was 2–3 m lower than that of today, i.e. 4,000–2,000 years ago (ISEKI, 1974).

6.5.3 Hokuriku and San-in

Deposits younger than Middle Pleistocene in the Hokuriku district consist of fluvial gravels of the high, middle and low terraces. On the other hand, around the Noto Peninsula coastal terraces of the S surface are developed, but the former shore lines vary in height due to the later block movement. Fragmental terraces in the Wakasa Bay and the Tango Peninsula areas are mostly correlated with the S surface. Holocene sand dunes are widely observed in the coastal areas of this district.

6.5.4 Northern Kyushu

A number of terraces are developed in the Oita Plain district. They are classified, from higher to lower as I to VII. Of them, III, IV and V are marine, and the terrace V is of the Oka Formation correlated with the Shimosueyoshi Formation. Deposits of paleo-sand dunes are distributed in the Genkai-nada coast area and are intercalated with the pyroclastic flows of Aso Volcano (Aso-4: 70,000–80,000 years old). The marine terraces of the S surface are also developed in the Ube district in the Setouchi region.

Around the Ariake and Shiranui inland seas the high, middle and low terraces are fragmentarily distributed. The middle terrace is covered by the Yame clay (=Aso-4 flows) and comprises the marine Nagasu Formation, which is correlated with the Shimosueyoshi Formation. Marine deposits representing the early and late stages of the Yurakucho Transgression are, respectively, the Shimabara Bay sand-gravel and the Ariake clay (ARIAKE-KAI RESEARCH GROUP, 1965).

6.6 Nansei-shoto

In the Okinawa main island, the Ryukyu Group (Middle and Upper Pleistocene) overlies the Shimajiri Group (Pliocene to Lower Pleistocene). Similar deposits are also distributed in the islands of Amami, Okinawa and Sakishima. The Ryukyu Group occurs on the coastal terraces lower than 200 m and consists mainly of reef limestone (Ryukyu Limestone). It contains corals, molluscs, foraminifers, bryozoans and algae, and subordinate conglomerate, sandstone, siltstone, tuff and lignite (NAKAGAWA, 1967).

The remains of a number of continental animals are obtained from this archipelago and suggest that the islands may have been connected with the Asian Continent.

The ^{230}Th, ^{231}Pa and ^{14}C age studies on the Ryukyu and Araki Limestones of island Kikai have given the following results: 0.13–0.10 Ma for the older, 90,000–80,000 years for the middle and 60,000–50,000 years for the younger Ryukyu Limestone, and 50,000–35,000 years for the Araki Limestone (Fig. 6.6; KONISHI *et al.*, 1974).

Outside the Ryukyu Limestone terraces are developed Holocene elevated terraces of reef limestone, 1–20 m in altitude. They show 8,000–1,000 ^{14}C years.

In the Osumi Islands, in the northernmost part of the archipelago, are developed high, middle and low terraces, all being lower than 300 m, and the highest of the middle group seems to be the S surface. The lower group terraces, about 10 m in altitude, may have been formed during the Holocene transgression.

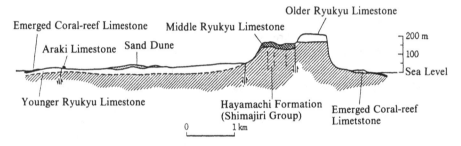

Fig. 6.6.   Geologic cross section of Kikai-jima (after KONISHI *et al.*, 1974).

REFEREMCES

Ariake Bay Research Group, Quaternary System of the Ariake and the Shiranui Bay areas, with special references to the Ariake Soft Clay, *Assoc. Geol. Collabor. Japan Monogr.*, **11**, 1–86, 1965.

HOKURIKU QUATERNARY RESEARCH GROUP, The Quaternary deposits an history in Hokuriku region, with reference to the eustatic and plaeoclimatic changes and the tectonic movements, *Quaternary System of Japan*, Research Group on the Quaternary of Japan, ed., 263–297, 1969.

HOSHINO, Y., Some problems on the geomorphic development in the Miyazaki plain, southeastern Kyushu, *Quatern. Res.*, **10**, 99–109, 1971.

ISEKI, H., Sea level toward 2,000 yr B.P. in Japan, *J. Fac. Liter., Nagoya Univ.*, **62**, 155–176, 1974.

ITIHARA, M., Some problems of the Quaternary sedimentaries, Osaka and Akashi areas, *Chikyu Kagaku (Earth Sciences)*, **49**, 15–25, 1960.

KAIZUKA, S., NARUSE, Y., and MATSUDA, I., Recent formations and their basal topography in and around Tokyo Bay, central Japan, *Quatern. Res.*, **8**, 32–50, 1977.

Kanto Quaternary Research Group, Some problems on the geology of the Shimosueyoshi upland, *Chikyu Kagaku (Earth Science)*, **24**, 151–166, 1970.

KONISHI, K., OMURA, K., and NAKAMICHI, O., Radiometric coral ages and sea level records from the late Quaternary reefr complexes of the Ryukyu islands, *Proc. 2nd Intn'l Coral Symp.*, **2**, 595–613, 1974.

MACHIDA, H., ARAI, F., MURATA, A., and HAKAMATA, K., Correlation and chronology of the Middle Pleistocene tephra layers in south Kanto, *J. Geogr.*, **83**, 302–338, 1974.

MINATO, M., FUJIWARA, Y., and KUMANO, S., Subsurface geology of the Late Quaternary deposits in Hokkaido, *Mem. Geol. Soc. Japan*, **7**, 1–11, 1972.

NAKAGAWA, H., Geology of Tokunishima, Okierabujima, Yoronto and Kikaijima, Amami Gunto, *Contrib. Inst. Geol. Paleontol., Tohoku Univ.*, **63**, 1–39, 1967.

NAKAGAWA, H., OGAWA, S., and SUZUKI, J., Quaternary geology and geomorphology of Sendai and its environs (1). *Quatern. Res.*, **1**, 219–227, 1960.

NOBI PLAIN QUATERNARY RESEARCH GROUP, Stratigraphy and microfossil analysis of Quaternary sediments in the Nobi plain, central Japan, Mem. Geol. Soc. Japan, **14**, 161–183, 1977.

Nojiri-ko Excavation Research Group, *The Lake Nojiri Excavation 1962–1973, Kyoritsu Shuppan Co.*, 278p., 1970.

OTA, Y., Late Quaternary vertical movement in Japan estimated from deformed shorelines, Quaternary Studies (R. P. Suggate and M. M. Creswell, eds.), *Roy. Soc. New Zealand Bull.*, **13**, 231–239, 1975.

SUZUKI, K., MANABE, K., and YOSHIDA, T., The Late Cenozoic stratigraphy and geologic development of the Aizu basin, Fukushima Prefecture, Japan, *Mem. Geol. Soc. Japan*, **14**, 17–44, 1977.

SUZUKI, K., YOSHIDA, T., ITO, S., and SOHMA, K., Quaternary history of the Koriyama basin, *Sci. Rept. Fac. Educ., Fukushima Univ.*, **17**, 49–67, 1967.

Toyono Collaborative Research Group, Quaternary geology and geological history in the western marginal area of the Nagano basin, *Mem. Geol. Soc. Japan*, **14**, 79–92, 1977.

TSUCHI, R., Geologic structure and history of "Udo" Hill in Shizuoka Prefecture, central Japan, *J. Geol. Soc. Japan*, **66**, 251–262, 1960.

YOSHIKAWA, T., KAIZUKA, S., and OTA, Y., Mode of crustal movement in the late Quaternary on the southeast coast of Shikoku, southwestern Japan, *Geogr. Rev. Japan*, **37**, 627–648, 1964.

CHAPTER 7

CENOZOIC VOLCANIC ACTIVITIES AND THEIR PRODUCTS

7.1 Temporal and Spatial Distribution of Cenozoic Volcanoes

7.1.1 Historical outline of Cenozoic volcanism

Considerable volcanism resumed in Late Eocene to Early Oligocene times in the Japanese Islands, since the cessation of the immense felsic igneous activity of late Cretaceous. The representative products are pyroxene andesite (39–41 Ma) of Haha-jima in the Bonin Islands, island-arc type basaltic rocks recovered at Site 448 of Leg 59 of DSDP, 17°N on the Kyushu-Palau arc, and the pillow lavas and sheets of alkali basalt in the Paleogene Kumage Group in the Yaku and Tane Islands.

In Northeast Japan the volcanic front is considered to have migrated westwards at the rate of 0.1 cm/year from Miocene to the present (MATSUDA and UYEDA, 1971; Fig. 7.1). Miocene volcanism was mostly submarine, and during the Early Miocene the erupted rocks were andesitic, while in Middle Miocene times, rhyolite, dacite and basalt were predominant. Due to the ubiquitous crystallization of chlorite, sericite and zeolite, the rocks are commonly altered into pale green rocks called "Green Tuffs" by Japanese geologists. Volcanic activities in this region declined during late Middle Miocene to Pliocene (Fig. 7.2).

In the same period the volcanic front of Southwest Japan moved more than 200 km southwards compared with that of the previous time. As a result, particular rocks such as Mg-rich andesite and biotite rhyolite with garnet phenocrysts erupted in the Setouchi province. Gigantic volcano-plutonic complexes of felsic composition, such as those of the Kumano and Osuzu districts were formed in the Pacific coast areas. Furthermore, tholeiitic and alkali basalts were extruded and intruded in the Sea of Japan coastal area extending from northern Kyushu to the San-in provinces.

7.1.2 Distribution of volcanoes in relation to subduction zones

About 200 Quaternary volcanoes are located on the Japanese Islands and are mostly considered to be of the consumed plate margin type, i.e. volcanoes of an island-arc. Two volcanic zones are classified: the East Japan and West Japan Volcanic Zones. The East Japan Volcanic Zone is related to

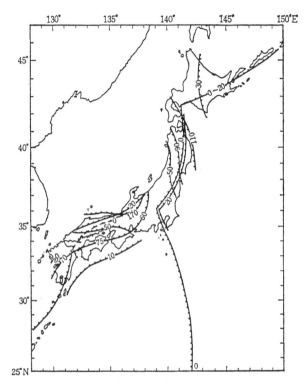

Fig. 7.1. Temporal variation of the volcanic front (MATSUDA and UYEDA, 1971). The respective volcanic zones were located on the sides, indicated by short bars.

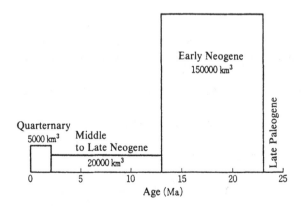

Fig. 7.2.   Estimated volumes of the volcanic ejecta since Neogene time (SUGIMURA, 1974).

subduction of the Pacific plate, and is subdivided into those on the Kurile, Northeast Japan and Izu-Mariana arcs. The West Japan Volcanic arc consists of the Southwest to Ryukyu arc, and corresponds to subduction of the Philippine Sea plate. However, exceptional features of the arc-volcanic zone are observed in the Japanese Quaternary volcanoes. Firstly, the active Benioff Zone just reaches the central part of Shikoku, but further to the north Quaternary volcanoes have been developed in the Sea of Japan coast area. Thus the volcanoes and the Benioff Zone do not relate directly to each other in Southwest Japan. Secondly, the distance between the trench and the volcanic front is not always invariable as shown in Fig. 7.3; it becomes much larger at the meeting point of the two arcs (ARAMAKI and UI, 1982).

In a few areas, well away from the volcanic front, volcanoes are densely aggregated. This is observed, for example, in the Ontake-Norikura-Tateyama volcanic zone in central Japan and in the volcanic clusters of the Shiretoko, Tokachi-Daisetsu-Tengu, Niijima-Kozushima, Hakone-Amagi and Tomurosan-Hakusan district. Some of the clusters stretch in an oblique direction to the front.

7.1.3   Stratovolcanoes, calderas and other features

Stratovolcano and caldera are the two major types of Japanese Quaternary volcanoes.

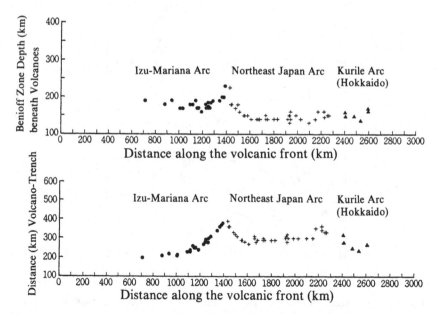

Fig. 7.3.   Relations between the depth of the Benioff zone, the distance of the volcano and the trench, and the distance of volcanoes along the front (after ARAMAKI and UI, 1978).

Stratovolcanoes are not infrequently built on a preceding volcano or volcanoes. An example is given by Mt. Fuji, which has been formed on two old volcanoes: Kofuji and Komitake. Where a stratovolcano is highly dissected, radial dike-swarms originating from the central conduit are exposed. The dikes are often bent to become parallel to the direction in which the parasitic volcanoes also are arranged. The configuration indicates the direction of the regional stress field at the time of magmatic activity (NAKAMURA, 1975; Fig. 7.4).

Recent studies on Japanese collapse calderas such as Kutcharo, Towada, Aso and Aira, indicate some features not in agreement with the model proposed by H. Williams in 1941. First, no gigantic stratovolcano would have previously existed, because we can find few exposures of volcanic rocks under pyroclastic flow deposits on caldera walls. Second, the amount of pyroclastic flow is not necessarily comparable to the dimension of the respective calderas. Four big pyroclastic flows, lasting for a few tens of thousand years, all originated from a single Aso caldera. Third, the drilling

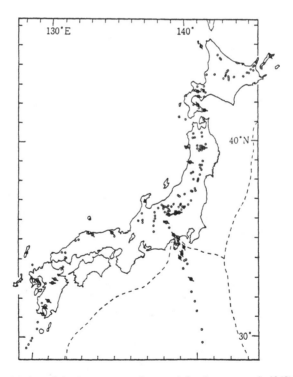

Fig. 7.4. Distribution of the Quaternary volcanoes (after ISSHIKI *et al.*, 1968), and trends of parasitic cones or monogenic volcanoes indicated by NAKAMURA, 1975. The trends roughly agree with the present-day maximum compressional axis in the Japanese Islands.

survey indicated that surfaces of the basement rocks are usually not so deep at the Hakone and Aso calderas (Fig. 7.5), and that highly brecciated zones involving the basement were detected in the central part of the Kutcharo caldera. The gravity anomaly also decreases gradually inwards.

Consequently, the following revised theory for Japanese collapse calderas is proposed (ARAMAKI, 1969; YOKOYAMA, 1969). The explosion of magma at shallow depth produces pyroclastic flows and causes brecciation of the existing rocks. A shallow depression is thereby formed and develops into a caldera. The process would resume at the same site, because a new small caldera is often observed within an old large caldera. In some cases, new volcanoes are built on the caldera edge, for example, Kamuinupuri on the Mashu and Sakurajima on the Aira. Finally, calderas are highly limited in distribution in the Hokkaido, northern Tohoku and Kyushu districts, and those larger than 3 km in diameter are not found in areas more than 30 km distant from the volcanic front.

Groups of small monogenetic volcanoes are developed, for example, in the Megata, Higashi-Izu, Abu, Kannabe, Goto Islands, Iki and Oki districts.

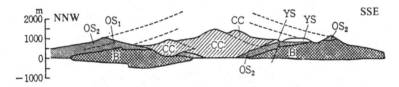

Fig. 7.5. Underground structure of the Hakone caldera inferred from by boring. B: basement rocks, OS_1 and OS_2: Strato-volcanoes before the old caldera, YS: Strato-volcano before the young caldera, CC: Central cones after the caldera formation.

7.1.4 Regional variation of chemical composition and amount of ejecta

KUNO (1960) discussed that the following three zones of Quaternary volcanoes, each characterized by distinct basalt types, run parallel to the direction of the Japanese islands arc. They are, from the volcanic front inwards (westwards),

1) Zone of tholeiite,
2) Zone of high-alumina basalt, and
3) Zone of alkali basalt.

Figure 7.6 shows the relationship between the SiO_2 content of volcanic rocks and their volume. The volume for a given SiO_2 content shows the sum of volumes obtained by dividing the volume of a volcano by the number of analyses (ARAMAKI and UI, 1978). The figure indicates that the SiO_2 content of the maximum volume varies from arc to arc. Basalt predominates in the

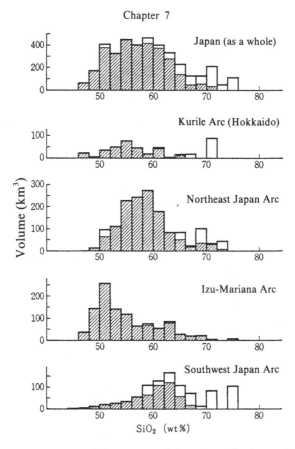

Fig. 7.6. Relations between the SiO₂ content and the volume of volcanoes in the respective volcanic zones (after ARAMAKI and UI, 1978). Hatched and blank areas respectively indicates lavas and pyroclastic rocks.

Izu-Mariana arc, while andesite predominates in the Northeast Japan arc. The second maximum observed in the Kurile, Northeast Japan and Southwest Japan arcs resulted from vast amounts of felsic pyroclastic materials.

The K-values defined by DICKINSON *et al.* (1967) are plotted on the map of Fig. 7.7. Generally, the isopleths are harmonious with the trend of the Benioff Zone. However, in the Northeast Japan arc the value gradually increases from the northern end southwards to Bandai: the along-arc variation is observed. The local high-K and low-K are found respectively in Tokachi, Yatsugatake and Aso Volcanoes and in Asama and Haruna Volcanoes. The anomalies occur in positive and negative relations, respectively, with the thickness of the crust, and with the long-range Bouguer anomaly (UI and ARAMAKI, 1978). The K_2O content of Japanese Quaternary

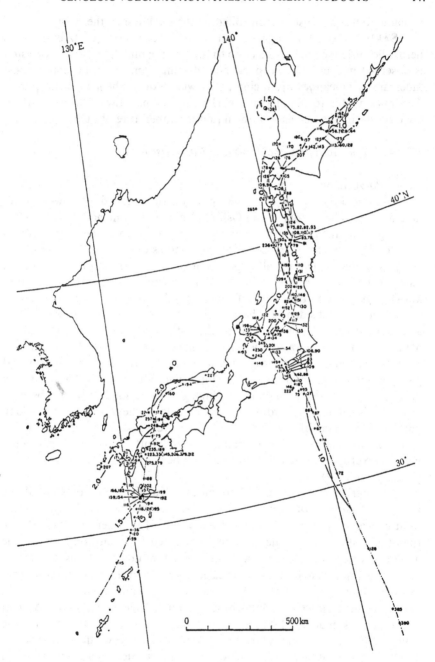

Fig. 7.7.   Dickinson's K^{60}. The figures for each volcano are 100 times the respective K-values, while contours are drawn on the basis of the K-value themselves (after ARAMAKI and UI, 1978).

volcanic rocks may also be controlled by the thickness of the crust.

SAKUYAMA (1977) suggested that volcanoes where quartz crystallizes before hornblende are located close to the front, while the reverse succession is observed in rocks of volcanoes more distant from the front, and in most inner areas, volcanoes producing rocks with biotite phenocryst are found. This may be due to the increase of H_2O in magmas inward from the front area, resulting in a decrease of the liquidus temperature of quartz.

7.2 Volcanic Activities and Products: Case Histories

7.2.1 Polygenetic volcanoes

Oshima Volcano is a typical polygenetic volcano of low-viscosity magma type. The history of old Oshima Volcano commenced about 20,000 years ago, and the main stratovolcano and parasitic volcanoes were built on the Late Pliocene to Pleistocene volcanic bodies of Okada, Fudeshima and Gyojanoiwaya, which are all exposed in the northern and eastern coastal parts of the island. About 1,500 years ago, a small Kilauea-type caldera (3 km×4 km, 160 m deep) was formed on the summit through phreatic explosion and subsidence. The caldera was once filled with water, and aquagene tuff was deposited. During the last 1,400 years, the central cone of Miharayama and parasitic volcanoes have been active. Thick falls of pyroclastic deposits were formed around the crater of Miharayama and partly welded so as to make a steep slope. The caldera was often filled with dense lavas. As a result, an unusually high gravity anomaly is today observed around this caldera. According to detailed field studies (NAKAMURA, 1964), ten cycles of deposition are detected. One cycle lasted 153 ± 50 years and consisted successively of scoria-fall, lava-flow, tuff layers and a weathered layer. The eruption rate is inferred to have been about 2.8×10^{-3} km^3 a year (Figs. 7.8 and 7.9).

The earlier stage of the stratovolcano, Kurofuyama, commenced the history of Asama Volcano (Fig. 7.10). The central part of the edifice was built of welded tuff breccia derived from falls of pyroxene andesite. Since then the eastern half of the mountain collapsed by explosion to make the Tsukahara debris avalanche. After this a few minor eruptions resulted in dacitic lava domes and dacitic pyroclastic flows, and then the major activity resumed. During the last several thousand years, a number of andesitic lavas and pryoclastic flows have built Maekakeyama Volcano to 6 km^3. A great eruption took place in 1783, and the Agatsuma and Kanbara pyroclastic flows and the Onioshidashi lava flow occurred for just two days (ARAMAKI, 1963). The Agatsuma pyroclastic flow is weakly welded, and many stretched hilly rises occur on its surface showing the development of the flow direction (Fig. 7.11). On the other hand, the Kanbara pyroclastic flow carries a

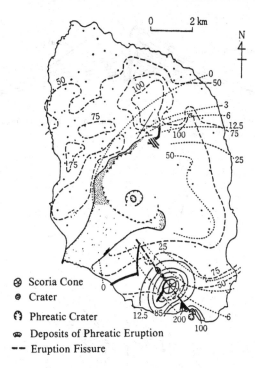

Fig. 7.8.   Ejecta, for one cycle of activity, of the Oshima volcano, (after NAKAMURA, 1964), showing ejecta in the period 1421–1552. Broken line: Isopack of pyroclastics ejected from the caldera crater. Dotted line: Thickness of the basal scoria beds (cm), Solid line: Isopack (cm) of pyroclastics ejected from parasitic fissures, Black: Lavas, Stippled: The area where ejecta were eroded away by wind, Dot: Determining point for the thickness of pyroclastics.

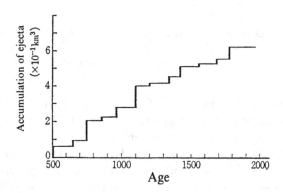

Fig. 7.9.   Accumulative curve of ash derived from the younger Oshima volcano (after NAKAMURA, 1964).

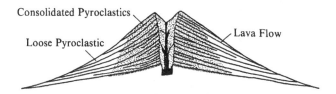

Fig. 7.10. Restored cross-section of Kurofuyama, the old strato-volcano of the Asama volcano (after ARAMAKI, 1963).

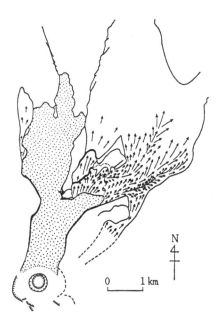

Fig. 7.11. The 1783 ejecta of the Asama volcano (after ARAMAKI, 1968b). A number of elongated hills carrying essential blocks are observed on the Agatsuma pyroclastic flow deposits; their trends follow pre-existing valleys. Stippled: Onioshidashi lava flow.

number of juvenile blocks, some of which reach 30 m in size (Fig. 7.12). When reaching a point about 8 km from the crater, the flow changed into a mud flow, having taken in surface waters.

The history of *Fuji Volcano* began about 80,000 years ago. It is divided into two stages: Kofuji (old Fuji) and Shin-Fuji (new Fuji). Kofuji erupted basaltic lavas and pyroclastics, and at the end of its history an explosive eruption of felsic magma occurred. After a pause, the basaltic eruptions of Shin-Fuji commenced which lasted until 1707, when an explosion of dacitic ejecta took place at a crater on the slope. The form of the modern Fuji

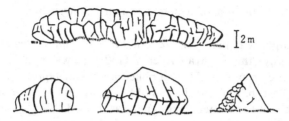

Fig. 7.12.   Gigantic essential blocks in the Kambara pyroclastic flow deposits (after ARAMAKI, 1968b).

Volcano is not perfectly conical but slightly elongated in the NW-SE direction, a direction in which many parasitic volcanoes are also arranged.

    Chokai Volcano lies in the Sea of Japan coastal area of Northeast Japan; it is a large (about 200 km³) andesitic stratovolcano. It has two central vents about 3 km distant from each other, while a number of parasitic craters are arranged in the N75°W direction (Fig. 7.13). On the western slope, a few normal faults have been developed, some of which cut the parasitic lava flows (UI, 1972). *Gassan Volcano* situated about 60 km due south of Chokai is also an andesitic stratovolcano, but at the closing stage of its history, a Peléean type eruption occurred. An explosion crater was formed in associa-

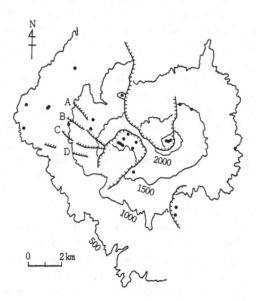

Fig. 7.13.   Parasitic cones (dots) and normal faults (A–D) of the Chokai volcano (after UI, 1972).

tion with the formation of a debris avalanche, pyroclastic flow and lava dome.

The complicated history of *Hakone Volcano* commenced about 0.4 Ma. Early in 0.2 m.y., the old stratovolcano (130 km^3) was built by basaltic and mafic andesitic eruptions. They supplied abundant tephra over the Kanto Plain to the east (p. 153). At the end of this phase, the old caldera (12 km×8 km), probably of the Crater Lake-type, was formed. Within the caldera, the new stratovolcano (14 km^3) had grown up through eruptions of pyroxene andesite and dacite. Furthermore, the new smaller caldera (7 km in diameter), of the Crater Lake-type, was formed in association with pyroclastic flows of more than 14 km^3. During the last 45,000 years, central cones have been developed. They are, for example, the andesitic lava domes of Komagatake and Futagoyama and the Kamiyama stratovolcano. Of notable recent events, the formation of the Kamiyama explosion crater (4,000 years) and of the Kammurigatake pyroclastic flows (2,000 years) due to a Peléean type eruption, should be mentioned here.

The principal vents of *Sakurajima Volcano*, Kyushu, have moved southwards from Kitadake through Nakadake to Minamidake. But the major eruptions usually occur at the parasitic craters opened up on the slope. At the time of the 1914 great eruption, 1.6 km^3 lavas and 0.6 km^3 pyroclastic falls issued forth for four months. The whole volume of Sakurajima Volcano is estimated to be 7.4 km^3, including 13 tephra layers which originated from them. On the other hand, the ^{14}C age data indicates that activity commenced about 11,000 years ago. Thus the eruption rate is inferred to be about 10^{-3} km^3 a year, and the great eruptions (about 1 km^3) are inferred to occur once in every 1,000 years.

Usu Volcano, Hokkaido, opened its history by the formation of a small stratovolcano on the southern edge of Toya caldera, which consists of basalt and mafic andesite. The last event of this stage was the Zenkoji debris avalanche about 7,000–8,000 years ago. After a quiescent period dacitic magmatism has resumed since 1663. A number of lava domes have been formed on the summit and on the northern to eastern slope. Some of the domes did not appear on the surface (Fig. 7.14; YOKOYAMA, 1973).

7.2.2 Pyroclastic deposits and calderas

The *Ito* pyroclastic flows erupted about 22,000 years ago from the Aira caldera of the Kagoshima Bay area in southern Kyushu. The eruption leading to the Osumi pumice fall and Tsumaya pyroclastic flow occurred first. Then in the northeastern part of the caldera a vent was opened, from which abundant essential and accidental ejecta were thrown out, rapidly flowed down and were deposited (ARAMAKI, 1969). The resultant Kamewarizaka breccia consists of intercalations of a number of fall and powder flow

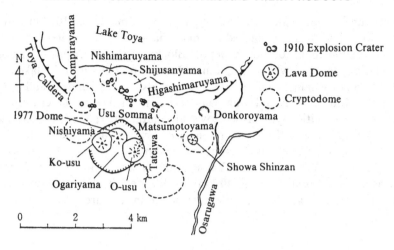

Fig. 7.14.   Lava domes of the Usu volcano (modified after YOKOYAMA *et al.*, 1973).

deposits. Subsequently, the main part of the Ito flows were formed. The Ito flows extend widely over southern Kyushu, and amount to about 150 km^3 and are only partially welded. The boundaries of the flow units are generally obscured. The Kyomachi Formation in the Kakuto basin district is made of graded pyroclastic deposits suggesting that a pyroclastic flow rushed into a lake that existed there at that time (ARAMAKI, 1968). The lower part is composed of pumice, pebble and sand, the upper part of silt and clay. Kyomachi Formation is correlated with the Ito pyroclastic flow using the characteristic higher refractive index of orthopyroxene. Recent study suggests that the Ito flows may not be related to the initial birth of Aira caldera, but are younger than the latter, because a part of the flows has been found on the erosion surface of the inner wall of the caldera (ARAMAKI, 1966).

The pyroclastic flows that originated from the *Aso* caldera are classified into four: from earlier to later, Aso 1, 2, 3 and 4, and large erosional gaps are formed between them. In the basal part of Aso 4 breccia formation similar to that of the Kamewarizaka breccia mentioned above is found (ONO, 1965).

The 1,000 m drilling in *Kutcharo* caldera, Hokkaido, showed that the recovered materials are mostly tuffaceous with little lava (only 2%) and low in density (1.5–2.2 g/cm^3). The caldera may be a shallow funnel-shaped depression resulting from an explosive eruption and filled with fall-back materials.

The history of the Miocene *Kumano* felsic rocks, extending over a vast area of 600 km^2, in the Kii Peninsula, started with the formation of a rhyolitic lava plateau that erupted from fissures. After a prolonged period, during which weathering proceeded a few meters depth into the surface, a great eruption took place. Vesiculated magmas of more than 100 km^3 issued forth for a few days and formed a funnel-shaped lava lake, larger than 400 km^2 in extent and 1,000 m in depth. Blocks of the basemental Kumano Formation, some of which are 2 km×10 km, dip at 30° into the lake (Fig. 7.15). Similar gigantic calderas are observed also in the Sobosan and Osuzu districts, Kyushu. The Sobosan caldera was formed after the eruption of dacitic to rhyolitic pyroclastic flows amounting to about 40 km^3, and attaining 12 km×16 km×850 m in size. On the other hand, the part remaining of the Osuzu caldera is about 200 km^3.

At the southern foot of Mt Yatsugatake in central Japan, the *Nirazaki* debris avalanche extends over 40 km. Its thickness is about 200 m, and its volume 9 km^3. Huge blocks, sometimes 500 m in size, are enveloped within

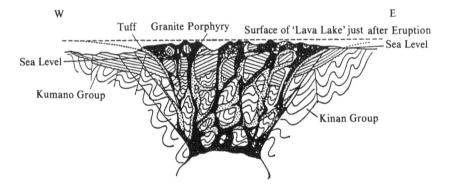

Fig. 7.15.   Restored cross-section of the Kumano felsic complex (after ARAMAKI, 1965).

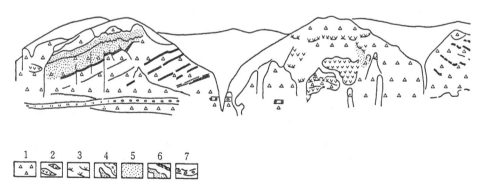

Fig. 7.16.   A sketch showing the internal structure of the Nirasaki pyroclastic flow (KOFU BASIN QUATERNARY RESEARCH GROUP, 1969). 1: Pyroclastics, 2: Autobrecciated lava, 3·4: Unbrecciated lava, 5: Reddish purple matrix, 6: Ash·Sand.

the flow; they retain the stratified structure of lava and pyroclastics (Fig. 7.16). A study of natural remanent magnetism indicates that the blocks have been derived from a stratovolcano. (KOFU BASIN RESEARCH GROUP, 1969)

Hummocky hills are found in the southern foothills area of the Shikaribetsu Volcano, Hokkaido. They are composed of the Higashi-Nupukaushi lava dome and poorly sorted juvenile blocks derived from it. Secondary fumaroles, carbonized wood blocks and surface oxidation indicate the high temperature existing at the time of sedimentation. The direction of natural remanent magnetism in the blocks is uniform.

Base surge deposits are observed in some monogenetic volcanic clusters of alkali basalt to trachyte such as those in the Goto Islands, San-in and southern Kyushu. Megata maars in Akita Prefecture is also the source of base surge deposits, which are well known in carrying a number of xenoliths derived from upper mantle and lower crust. Base surge deposits of rhyolitic composition are reported, for example, from Niijima Island, Izu.

7.2.3 Subaqueous volcanic sediments

The Wadaira Tuff Formation of the Miocene Tokiwa Group in the southern Fossa Magna region is composed of five tuff beds, along with mudstone layers between them. The lower part of each tuff bed consists of vaguely graded, non-stratified dacitic lapilli tuff (0–50 m), while the upper part consists of laminated tuff of 10–20 m in thickness (Fig. 7.17). Each layer of the upper part shows a graded structure. The grain size of the clasts in general decreases upwards through the whole horizon. Thus, each tuff bed has a doubly graded structure, whose origin is interpreted as follows: A subaqueous eruption occurred and relatively coarse clasts were deposited

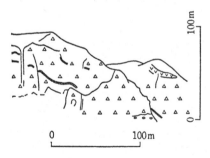

Fig. 7.16. (continued.)

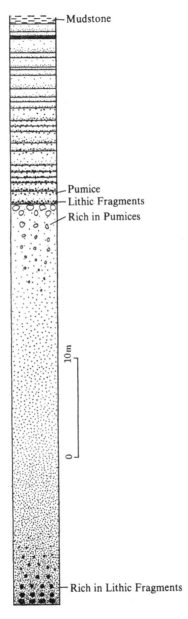

Fig. 7.17. Idealized columnar section of aquagene pyroclastic flow deposit (after FISKE and MATSUDA, 1964).

first, leading to rough grading of the lower part. The suspended fine particles were not quickly settled, but the occurrence of frequent turbidity currents led to the grading of thin layers in the upper part (FISKE and MATSUDA, 1964).

Basaltic pillow lavas have been frequently reported in Cenozoic volcanic rocks of Japan, for example, from the Oshoro district, Hokkaido, and the Kanzawa area in Fukushima Prefecture. They are generally both underlain and overlain by volcanic breccias. Andesitic hyaloclastite is described in the Irozaki area of Izu Peninsula. Furthermore, a number of dolerite sheet groups are observed, particularly in the Sea of Japan coastal area of northeast Japan.

7.2.4 Xenoliths

Lavas and pyroclastics of some Japanese Cenozoic volcanoes yield abundant xenoliths derived from lower crust and upper mantle. On the basis of detailed studies, a petrologic model for the crust and upper mantle in northeast Japan is proposed, as shown in Fig. 7.18 (TAKAHASHI, 1978).

7.3 Quaternary Tephra

About 40% in area of the Japanese Islands are covered by Quaternary tephra, whose total volume, at least on land, is estimated to be about 1,000 km^3. The distribution of the representative tephra is illustrated in Fig. 7.19. They are distributed generally in elliptical areas extending from the respective volcanoes, which are situated near the (western) ends of the long axes. The ^{14}C and fission-track ages of the fallout tephra have been measured down to about 0.4 Ma. Of the Japanese Quaternary tephra, Kanto Loam and Aira-Tn ash are described below.

7.3.1 Kanto Loam

Kanto Loam is a Middle to Late Pleistocene tephra deposit covering wide areas of the Kanto Plain, excluding the alluvial lowlands. In the central part of the plain it consists of highly weathered brownish loamy volcanic ash layers and a few seams of pumice, but in the marginal hilly and mountainous regions, it is comprised of various kinds of tephra such as scoria, pumice and ash. On the basis of terrace analysis and tephrochronology, Kanto Loam in southern Kanto Plain is classified into the following four formations. They are, from older to younger, the Tama, Shimosueyoshi, Musashino and Tachikawa Loams (KANTO LOAM RESEARCH GROUP, 1965; Fig. 7.20). The total thickness of the Kanto Loam is about 40 m in the Tama Hills district, but increases rapidly westwards to more than 200 m in the Oiso Hills.

Tachikawa Loam: Derived from the Kofuji Volcano, 10,000– 30,000 years ago. Reddish brown scoriaceous tuff rich in olivine. Intercalated with

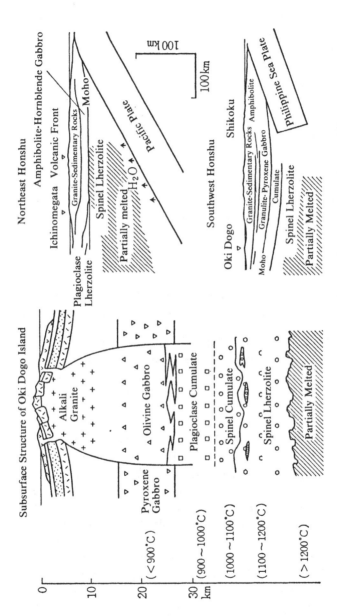

Fig. 7.18. Petrologic model for the crust and upper mantle in three regions of Japan, inferred from various xenoliths captured by alkali basalts (after TAKAHASHI, 1978).

four buried soil layers and one scoria seam. Often includes remains of Paleolithic age.

Musashino Loam: Brownish basaltic tuff derived from Kofuji Volcano, 30,000–60,000 years ago. In the lower part layers of the Tokyo Pumice (TP; 49,000 years) and the Miura Pumice (MP) has been intercalated. They are both related to the pumice flows that issued from the younger Hakone caldera.

Shimosueyoshi Loam: 60,000–130,000 years old tephra formation consisting of highly clayey tuff and 14 pumice layers. The upper part consists of olivine basalt from Fuji Volcano, while the middle to lower part consists of pyroxene andesite derived from the younger Hakone Volcano. In the lowermost horizon of the upper part the First Ontake Pumice (Pm-I; about 80,000 years) is interlayered.

Tama Loam: Thick brownish clayey tuff with more than 50 layers of pumice. 130,000–400,000 years old. In the Oiso Hills the materials are pyroxene andesite derived from the older Hakone Volcano, while in the areas to the north of the Tama Hills they are rich in biotite and hornblende which may have come from Yatsugatake Volcano.

Tephra deposits similar to those of the Kanto Plain are found in other areas and named after locality, for example, the Shinshu, Hachinohe, Daisen Loams and so on.

7.3.2 Aira-Tn Ash

A layer rich in volcanic glass fragments occurs in the middle part of the Tachikawa Loam. It has been utilized as one of age markers (22,000 years old). However the nature of the constituent minerals and particularly of the glass fragments, are quite different from all the other age index tephra of the Kanto region. Recently, it has been established that the volcanic glass layer represents the co-ignimbrite ash related to the Ito pyroclastic flows, which erupted from the Aira caldera in southern Kyushu. It is much further than previously supposed from the provenance locality (MACHIDA and ARAI, 1976). It is called Aira-Tn (AT) Ash. As shown in Figs. 7.19 and 7.21, it covers a vast area ($>2.7 \times 10^6$ km^2) extending from Kyushu to northern Honshu and the Sea of Japan floor. Its volume is estimated to be 300 km^3. The age of the Aira-Tn Ash is 22,000 years which corresponds to the peak of the Würm Ice Age.

The Akahoya tuff (Ah) of the Kikai caldera, located due south of Kyushu, is a similar co-ignimbrite ash layer. It is 6,300 years old, and covers about 1.4×10^6 km^2 and its volume amounts to 150 km^3 (MACHIDA and ARAI, 1983).

These co-ignimbrite ash deposits are also highly important in archaeology, because they are often intercalated with layers yielding artifacts.

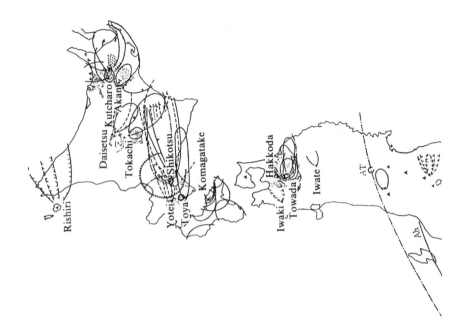

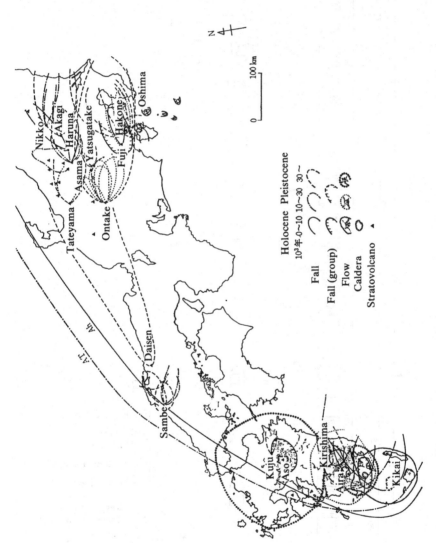

Fig. 7.19.   Quaternary tephra around the Japanese Islands (compiled after MACHIDA, 1976).

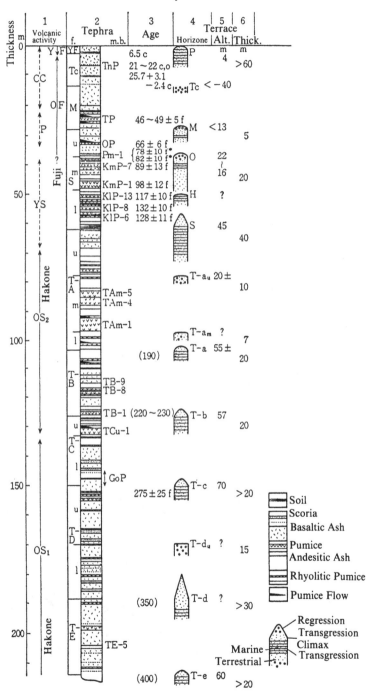

Fig. 7.20.

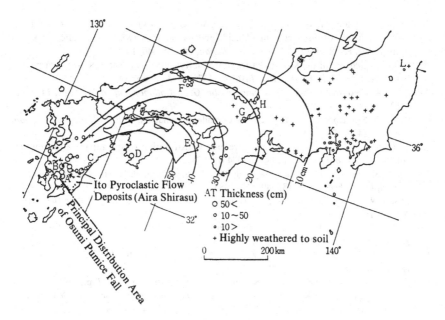

Fig. 7.21.   Distribution of the Aira Tn-ash (after MACHIDA and ARAI, 1976). A: Kamewarizaka, B: Shirokurabaru, C: Kawaminami, D: Sukumo, E: Kamikatsu-machi, F: Hirusembara, G: Ohara, Kyoto, H: Mihama, I: Hakone pass, J: Shidota, K: Aone, L: Kuwaori.

Fig. 7.20.   Kanto Loams and their tephrochronology in southern Kanto (after MACHIDA, 1975). 1. Volcanic activity CC: Central cones, P: Pumice flows, Y: Younger somma, OS_2 and OS_1: Old sommas, YF: Younger Fuji volcano, OF: Older Fuji volcano. 2. Tephra f: formation, m.b.: Marker bed, TnP, TP etc.: Tephra symbols. 3. Age ($\times 10^2$ years ago) c: ^{14}C, o: Hydration of obsidian, f: Fission-track (obsidian), f: Fission-track (zircon). 4. Stratigraphy P: Post-glacial, Tc:Tachikawa, M: Musashino, O: Obaradai, H: Hikihashi, S: Shimosueyoshi, T: Tama. 5. Altitude   Altitude (m) of the paleo-shore line in the northern Yokohama area. 6. Maximum thickness of sediments observed in the field.

REFERENCES

ARAMAKI, S., Geology of Asama Volcano, *J. Fac. Sci., Univ. Tokyo, Sec. II*, **14**, 229–443, 1963.

ARAMAKI, S., Mode of emplacement of acid igneous complex (Kumano Acidic Rocks) in southeastern Kii Peninsula, *J. Geol. Soc. Japan*, **71**, 525–540, 1965.

ARAMAKI, S., Geology of the Kakuto basin, southern Kyushu, and the earthquake swarm from February, 1968, *Bull. Earthq. Res. Inst.*, **46**, 1325–1343, 1968a.

ARAMAKI, S., Geology of Asama Volcano, *Assoc. Geol. Collabor. Japan Monogr.*, **14**, 1–45, 1968b.

ARAMAKI, S., Some problems of the theory of caldera formation, *Bull. Volcanol. Soc. Japan, 2nd Ser.*, **14**, 55–76, 1969.

ARAMAKI, S. and UI, T., Major element frequency distribution of the Japanese Quaternary volcanic rocks, *Bull. Volcanol.*, **41**, 390–407, 1978.

ARAMAKI, S. and UI, T., Alkali mapping of the Japanese Quaternary volcanic rocks, *J. Volcanol. Geotherm Res.*, **18**, 549–560, 1983.

DICKINSON, W. R. and HATHERTON, T., Andesite volcanism and seismicity around the Pacific, *Nature*, **157**, 801–803, 1967.

FISKE, R. S. and MATSUDA, T., Submarine equivalents of ash flows in the Tokiwa Formation, Japan, *Amer. J. Sci.*, **262**, 76–106, 1964.

ISSHIKI, N., M″2ATSUI, K., and ONO, K., Volcanoes of Japan, 1:2,000,000 Map Series No. 11, Geological Survey of Japan, 1968.

KANTO LOAM RESEARCH GROUP, *The Kanto Loam*, Tsukiji-shokan Co., 378p., 1965.

KOFU BASIN QUATERNARY RESEARCH GROUP, Geology of the southern foot of the Yatsugatake volcanic chain, central Japan, *J. Geol. Soc. Japan*, **75**, 401–416, 1969.

KUNO, H., High-alumina basalt, *J. Petrol.*, **1**, 121–145, 1960.

KUNO, H., OKI, Y., OGINO, K., and HIROTA, S., Structure of Hakone caldera as revealed by drilling, *Bull. Volcanol.*, **34**, 713–725, 1970.

MACHIDA, H., Pleistocene sea level of south Kanto, Japan, analysed by tephrochronology, Quaternary Studies (R. P. Suggate and M. M. Cresswell, eds.), *Roy. Soc. New Zealand Bull.*, **13**, 215–222, 1975.

MACHIDA, H., Stratigraphy and chronology of Late Quaternary marker tephras in Japan, *Geogr. Rept. Tokyo Metropol. Univ.*, **11**, 109–132, 1976.

MACHIDA, H. and ARAI, F., The widespread tephra—the Aira-Tn ash, *Kagaku (Science)*, **46**, 339–347, 1976.

MACHIDA, H. and ARAI, F., Extensive ash falls in and around the Sea of Japan from large Late Quaternary eruptions, *J. Volcanol. Geotherm. Res.*, **18**, 151–164, 1983.

MATSUDA, T. and UYEDA, S., On the Pacific-type orogeny and its model: extension of the paired belts concept and possible origin of marginal seas, *Tectonophys.*, **11**, 5–27, 1971.

NAKAMURA, K., Volcano-stratigraphic study of Oshima Voicano, *Izu. Bull. Earthq. Res. Inst.*, **42**, 649–728, 1964.

NAKAMURA, K., Volcano structure and possible mechanical correlation between volcanic eruptions and earthquakes, *Bull. Volcanol. Soc. Japan, 2nd Ser.*, **20**, 229–240, 1975.

ONO, K., Geology of the eastern part of Aso caldera, central Kyushu, southwest Japan, *J. Geol. Soc. Japan*, **71**, 541–553, 1965.

SAKUYAMA, M., Lateral variation of H_2O contents in Quaternary magma of northeastern Japan, *Bull. Volcanol. Soc. Japan, 2nd Ser.*, **22**, 263–271, 1977.

SUGIMURA, A., *Island Arcs, Physics of the Earth*, Physical Society of Japan, 190–222, 1974.

TAKAHASHI, E., Petrological model of the upper mantle and the lower crust of the island arc: petrology of mafic and ultramafic xenoliths in Cenozoic alkali basalt of the Oki-Dogo island in the Japan sea, Intern. Geodynamics Conf., Tokyo 1978, 334–335, 1978.

UI, T., Fault scarps on the slope of the Chokai volcano and genesis of pyroclastic rocks distributed at the flank, Chokai-san Tobishima, Yamagata Prefectural Office, 8–13, 1972.

YOKOYAMA, I., Some remarks on calderas, *Bull. Volcanol. Soc. Japan. 2nd Ser.*, **14**, 77–83, 1969.

YOKOYAMA, I. KATSUI, Y., OBA, Y., and EHARA, S., Komagatake, its volvanic geology, history of eruption, present state of activity and prevention of disaster, Committee for Prevention of Disaster of Hokkaido, Sapporo, 254p., 1973.

CHAPTER 8

SUBMARINE TOPOGRAPHY AND GEOLOGY AROUND THE JAPANESE ISLANDS

8.1 Continental Margin around the Japanese Islands (Fig. 8.1)

Geologic features commonly observed on the continental shelf such as ria coast, ancient shore-line sediments, drowned valleys, buried shelf channels and submarine terraces suggest that the sea-level was lower in ancient time. On the other hand, the greatest sea-level lowering of the Ice Age is indicated by a shelf break that occurs uniformly at a depth of about -140 m all over the world.

In some places around the Japanese Islands, the shelf break is deeper than -140 m, for example, in areas off San-in and Sanriku. This would have resulted from the crustal subsidence superimposed for a prolonged time on the glacial sea-level change. Beneath the shelf sediments unconformities are usually found. This is an effect of the crustal warping movement taking place since the Neogene (SATO, 1979).

Deep-sea plains are areas of turbidite deposits which fill the submarine tectonic basins. The structure of sediments, deduced from continuous seismic profiling, (c.s.p.) indicates that the development of basins and subsequent sedimentation, may have taken place subaqueously. The continental slopes are partially covered by deep-sea plain sediments. They are also incised by many submarine canyons, which suggests that the submarine canyons may also have been developed subaqueously. The submarine canyons often continue to channels on the abyssal plains (SATO, 1973).

A well developed continental borderland is observed in the Sea of Japan off Hokkaido to Noto Peninsula. Various geomorphic forms such as reefs, banks and basins are developed along the margin.

8.1.1 Sea of Japan margin (Figs. 8.2 and 8.7)

Along the continental margin, to the north of the Toyama trough in the Sea of Japan, a few highs and depressions are developed. Representative of

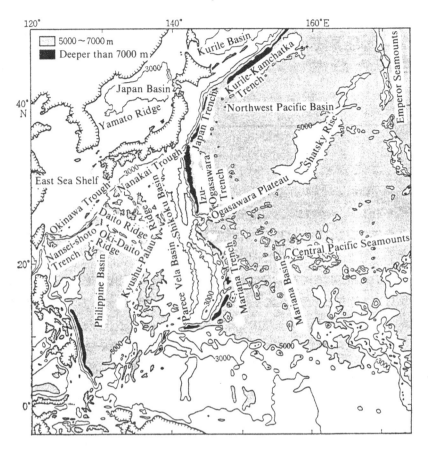

Fig. 8.1.   Outline of the submarine topography of Northwest Pacific.

these are the Okushiri Ridge off Hokkaido, and Sado Ridge, to the north of the Sado island. To the west of the Noto Peninsula, the shelf becomes wider and the shelf break deeper, and the submarine topography is dissimilar to that of the northern area mentioned above.

i)   *Western margin off Hokkaido*

The continuous seismic profiles, based on the detailed studies by SATO *et al.* (1973) and GEOLOGICAL SURVEY OF JAPAN (1978) are illustrated in Figs. 8.2 and 8.3. The width of the continental shelf is about 67 km off the Teshio River and 55 km in the area of Ishikari Bay; the shelf break occurrs at −140 to −160 m. An anticlinal structure (the Rishiri-Shokambetsu Upheaval Zone) is observed along the break from Rishiri Island to off-shore Mashike. It forms an embayment with a synclinal basin inside. The basin is filled by

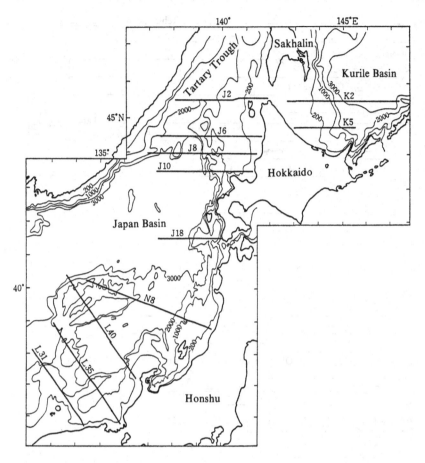

Fig. 8.2. Submarine topography of the Sea of Japan. Numbered lines indicate the paths of sonic survey (GEOL. SURV. JAPAN, 1978b, 1979).

thick sediments, and a basement high occurs in Ishikari Bay (Fig. 8.4).

To the west of the shelf break, several banks such as Musashi Tai (bank) and Otaru Tai were developed. They are all covered by a thin veneer of sediments comprising welded-tuff of probable pre-Neogene age, on the Musashi Tai and late Miocene siltstone on the Otaru Tai (GEOLOGICAL SURVEY OF JAPAN, 1978; SATO *et al.*, 1973).

The Okushiri Ridge on the western edge of the continental margin is a tilted block of sandstone and siltstone covered with younger sediments. Furthermore, sandstone and pumiceous siltstone are obtained from Shakotan Tai, and sandy siltstone and tuff breccia from the Kamui Submarine Hills. The rocks are considered to be of Miocene age. Thus the Neogene "Green

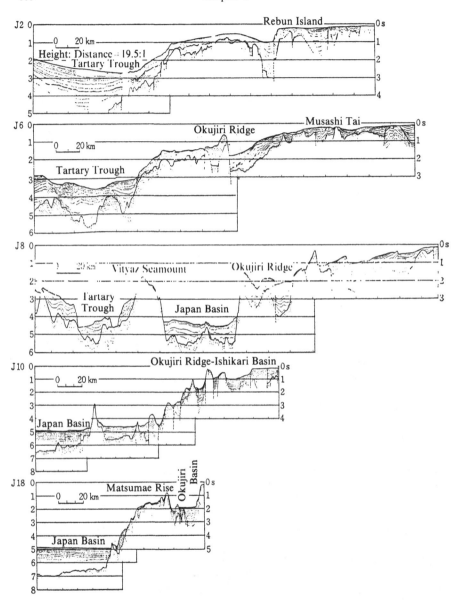

Fig. 8.3.   Cross-section of the sea-floor western off Hokkaido. J2 and others correspond to those in Fig. 8.2 (after GEOL. SURV. JAPAN, 1978b, 1979).

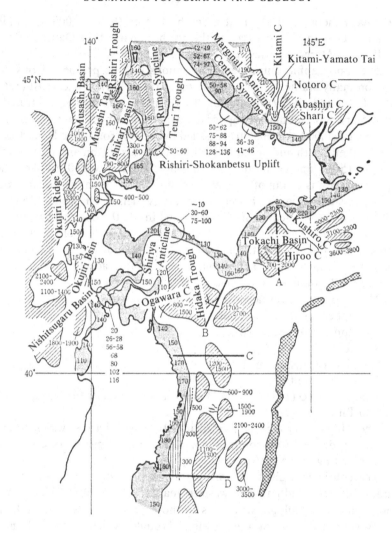

Fig. 8.4. Topographic division of the sea floor (Hokkaido-Tohoku). Gray: Shelf; figures indicate the depths of the shelf breaks (m), Vertical lines: Deeper shelf; figures indicate the depths of the breaks (m), Diagonal lines: Deep-sea plain and oceanic basin; figures indicate the depths (m), Diagonal cross pattern: Trench, c: Canyon, Lines A to D: measuring paths.

Tuff" region seems to extend along the continental margin from the Ishikari Bay to Musashi Tai.

On the continental slope to the west of the Tsugaru Peninsula Upper Miocene to Pliocene strata are distributed. Two submarine canyons on the slope reach the abyssal plain, at a depth of more than −3,000 m. Similar submarine canyons are observed in the northern part of the Okushiri Basin.

ii)   *Western Margin off Northeast Japan (Figs. 8.5 and 8.6)*

The topographic outline and continuous seismic profiles of this area based on many studies, such as those by HOTTA (1967), IWABUCHI (1968), SATO (1971) and GEOLOGICAL SURVEY OF JAPAN (1979) are illustrated in Figs. 8.5 and 8.6. Submarine topographic maps of almost the whole area on a scale of 1:200,000 or 1:50,000 are issued by the Hydrographic Department.

The shelf width and depth of shelf break in this region are, respectively, less than 15 km and −100 to −160 m (mostly −130 to 150 m). The shelf width, however, is 28 km off Niigata and only 2 to 3 km in the Toyama Bay. Where the shelf is comparatively wide, a basement high occurs along the shelf edge, and thick sediments are deposited inside the basement high.

The northern part of the Sado Ridge consists of a tilted block, while the southern part forms a horst. The rocks are of Neogene age and include, for example, volcanics near Tobishima island, diatomaceous or tuffaceous mudstone in Hyotan Se, and black mudstone in Kama Sho.

The c.s.p. survey revealed the following five layers:

1)   Upper acoustically opaque (A),
2)   Upper transparent (B),
3)   Lower opaque (C),
4)   Lower transparent (D), and
5)   Transparent (E).

According to SATO (1971), layers D and E are of Upper Miocene age and B and C of Pliocene age.

The Mogami Trough is divided into three sub-basins, which become successively deeper from south to north. Recent sediments have filled the basins, resulting in the flat surfaces.

The continental shelf and margin areas, extending from the western margin, off Hokkaido to the western margin, off Tohoku, have similar topographic and geologic structures to those on land in Northeast Japan. However, to the west of the Toyama Trough, which is located in the northeastern extension of Fossa Magna, the structures are different from those explained above. The Toyama Deep-sea Channel, has incised through the floor of the Toyama Trough. It is a valley which resulted from submarine erosion (IWABUCHI, 1968; TAGUCHI *et al.*, 1973). At the boundary between the Toyama and Yamato Troughs, a deep-sea fan has been developed, which contains meandering channels.

iii)   *Northern Margin off Southwest Japan (Figs. 8.5 and 8.6)*

To the west of the Noto Peninsula, the shelf break is anomalously deep, −300 to −500 m (MOGI, 1953). But the younger and shallower terrace is found at −140 m on the shelf (SATO, 1970). The latter is further divided into

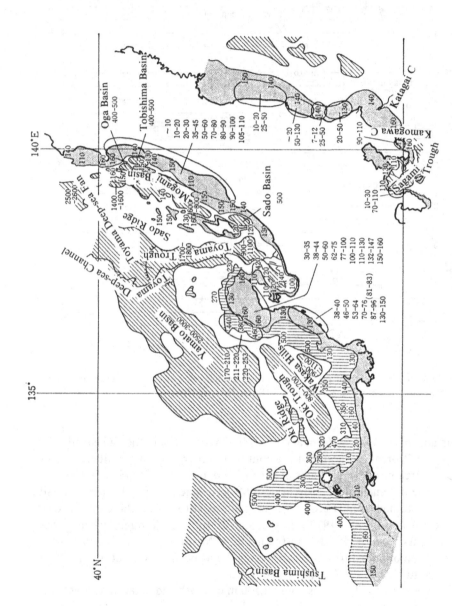

Fig. 8.5. Topographic division of the sea-floor of the Sea of Japan, and off-shore Joban and Boso. Symbols are the same as those in Fig. 8.4.

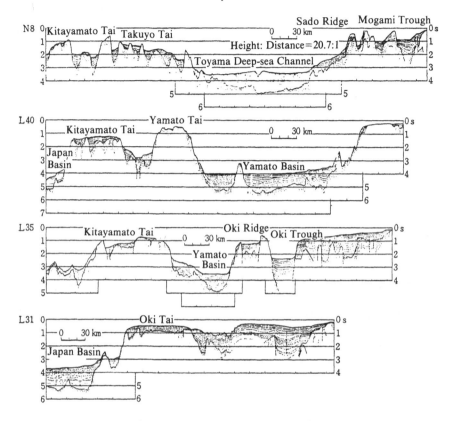

Fig. 8.6. Cross-section of the sea floor in the southern Sea of Japan area. N8 and others correspond to those in Fig. 8.2 (after GEOL. SURV. JAPAN, 1979).

surfaces of -150 to -160 m and of -110 m to the west of the Oki island. The deep shelf break suggest that the continental margin in the southern Sea of Japan area has generally subsided, causing deposition of thick sediments.

The basement is of early Middle Miocene age, being indicated by fossil shells obtained from O-guri to the west of the Noto Peninsula (GEOLOGICAL SURVEY OF JAPAN, 1979). It extends to the Oki Trough, where thick sediments cover it. Further to the north of Mishima island, Pliocene shells have been obtained from strata covering the Miocene basement rocks, which extend further to the Tsushima Trough area.

The Yamato Trough is about 2,500 m in depth and underlain by oceanic crust of about 14 km in thickness. It is filled with turbiditic sediments (HOTTA, 1967; GEOLOGICAL SURVEY OF JAPAN, 1979; KOIZUMI, 1977).

In the San-in district thick strata of Middle to Late Miocene are

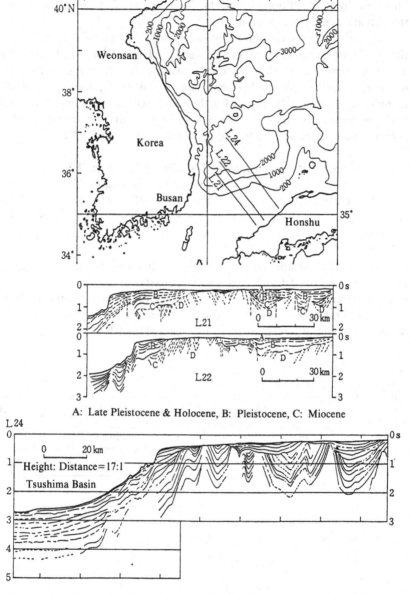

A: Late Pleistocene & Holocene, B: Pleistocene, C: Miocene

Fig. 8.7. Submarine topography and cross-sections in the southwestern Sea of Japan area (after GEOL. SURV. JAPAN, 1978a; OKAMOTO and HONZA, 1978).

developed on land in the coastal area, while the Oki Trough off San-in is underlain by thick Pliocene deposits. However, structural features between the two areas are not so variant.

In the Tsushima Trough to the west of the Oki Islands, fault basins in the pre-Cenozoic basement rocks are filled with Upper Oligocene to lowermost Miocene terrestrial to shallow-marine sediments. These are in their turn overlain by Early to Middle Miocene strata, consisting of deep-sea mudstone, turbidite and pyroclastic rocks. Furthermore, sediments younger than Late Miocene are terrestrial to littoral sandstone-mudstone (Upper Miocene) and shallow-sea sediments (Pleistocene to Recent) (MINAMI, 1979).

8.1.2   Off the coast of the Okhotsk Sea (Figs. 8.2, 8.4 and 8.8)

Sonic profiles across the Okhotsk Sea are given in Fig. 8.8 (NAGANO et al., 1974; GEOLOGICAL SURVEY OF JAPAN, 1978). To the north of Cape Notoro, the shelf is fairly wide and the break is comparatively deep, −170 to −210 m, while to the southeast it becomes narrower and shallower, −140 to −150 m.

The shelf to the north of Cape Notoro is divided into the outer marginal ridge, the inner shelf basin part and the coastal area. The outer marginal ridge consists of Jurassic basement rocks, the inner shelf basin is underlain by thick (4 km) sediments of Miocene to Pleistocene age, and the coastal area again consists of pre-Tertiary rocks (YAMAMOTO, 1979).

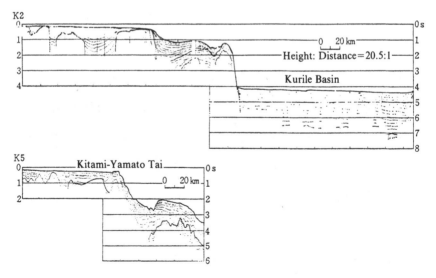

Fig. 8.8.   Cross-section in areas northeastern off-shore Hokkaido. K2 and K5 correspond to those in Fig. 8.2 (GEOL. SURV. JAPAN, 1978a).

Kitami-Yamato Tai is of a horst-like structure, whose western and eastern flanks are defined respectively by the Kitami and Notoro submarine canyons. A trough to the east of Kitami-Yamato Tai may extend into the Kurile Basin (YAMAMOTO, 1979).

8.1.3 Pacific margin

i) *Southern margin off Hokkaido (Figs. 8.4 and 8.9)*

The marine geology of this region has been studied by SATO (1962, 1970), UTASHIRO and SATO (1969), SAKURAI *et al.* (1975) and GEOLOGICAL SURVEY OF JAPAN (1977, 1978).

The shelf break is at -140 m depth off Cape Esan and -120 m depth off Shiraoi, but becomes deeper southwestwards to the Cape Erimo area, where it reaches -160 to -170 m. From Tokachi to Kushiro the shelf break is at -140 to -130 m, but off Akkeshi it becomes deeper, -210 to -220 m, and near the Nemuro Peninsula shallower again, -140 to -130 m. Off the

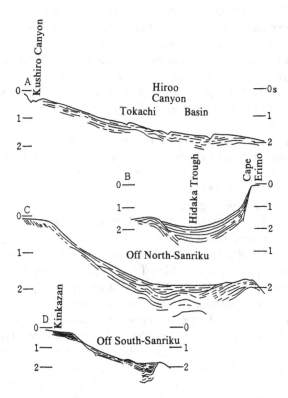

Fig. 8.9. Cross-sections in areas off Tokachi and Sanriku. A, C and D correspond to those in Fig. 8.4.

Yufutsu Plain a submerged beach ridge is found, which is traced on land to the inner part of the plain. It is suggested that the effect of tilting movement in this area surpressed the effects of sea-level lowering, after the Jomon Transgression (MOGI, 1964).

Figure 8.9 shows a few profiles across the sea floor, along the Pacific coast from off Tokachi, southwards to off Sanriku. In the Tokachi Basin, the tilted basement rocks are covered by sediments, in which a few short canyons are carved. The Hidaka Trough is filled with thick sediment accumulations, which form a synclinal structure. To the east of Muroran, a great arcuate collapse occurs. Off Ogawara Lake of Aomori Prefecture, a number of submarine canyons have been developed, some of which are buried by sediments, and at the base of the continental slope deep-sea fan have been constructed. (SAKURAI et al., 1975). The Hidaka Trough would seem to be a large turbidite-filled basin. The Shiriya Ridge which stretches northwards from Cape Shiriya is an anticlinal structure.

ii)   *Margin off Sanriku (Figs. 8.4 and 8.9)*

The shelf off Sanriku is fairly deep, -300 to -500 m, and it is shallower in the southern part, where a submarine terrace, at -150 to 180 m deep, has been developed. On the other hand, the shelf itself also becomes shallower in the northern part and coincides with the terrace surface in depth. Beneath the sediments covering the shelf, there is an unconformity, -250 to -300 m deep (NAKAJIMA 1973). The basement rocks of Middle Miocene age were dredged from an area off the southern part of the Shimokita Peninsula (SAKURAI et al., 1975).

Offshore from Sanriku, several deep-sea plains extend seawards from the outer ridge. Samples of diatomaceous mudstone of Late Miocene to Early Pliocene age were obtained from the deep-sea plain, at $-2,300$ m depth, offshore from Kinkazan.

iii) *Margin off Joban (Fig. 8.5)*

MOGI and IWABUCHI (1961) studied the continental shelf of this region, and areas deeper than the continental slope have been surveyed by MURAUCHI and ASANUMA (1970), and by GEOLOGICAL SURVEY OF JAPAN (1977). The shelf is 55 km in maximum width off Sendai, and its shelf break occurs at -130 to -150 m. Terraces -20 to -50 m deep are widely developed.

Similarly, as in the area off Sanriku, some deep-sea plains are observed. They are turbidite-filled basins underlain by synclinal basement structures. The axes of folding are roughly parallel to that of the Japan Trench, though in a few areas off-shore from southern Joban the axes cross each other.

The continental margin area of Northeast Japan consists of three parts: the tilted uplifted zone extending from the shelf to the slope, the uplifted zone of the outer part of the deep-sea plain, and the intervening subsiding zone, (HONZA, 1976a).

The submarine stratigraphy of this area was studied during an oil prospecting survey (KOMATSU, 1979), and is similar to that found off Sanriku. The strata are classified into four parts, from older to younger: (1) Upper Cretaceous, (2) Paleogene, (3) Miocene and (4) post-Miocene.

iv) *Margin offshore Boso and Sagami (Fig. 8.5)*

The shelf edge off Boso lies at -140 to -160 m depth. Distinctive submarine canyons cutting the shelf edge have been developed, and Miocene sediments are found to occur on the wall of the Kamogawa Canyon (HOSHINO and SATO, 1960). In the Sagami Bay, the eastern shelf is 5 to 8 km in width and the western shelf 2 to 5 km in width. The shelf break is at -110 to -130 m (MOGI, 1955).

According to studies by GEOLOGICAL SURVEY OF JAPAN (1975a), and KIMURA (1976), the Oligocene to Miocene strata (correlated with the Hayama Group or Mineoka Group) are exposed on the southward slope of the Boso Peninsula, the Lower Miocene (correlated with the Yugashima Group) to the east of the Izu Peninsula, and the Pleistocene on the floor of Sagami Bay.

A ridge and trough zone is developed on the northward wall of the Sagami Trough, east of its junction with the Kamogawa Canyon (IWABUCHI *et al.*, 1976). The wall formed in relation with the Sagami Tectonic Line, a right-lateral reverse fault. (KIMURA, 1976).

v) *Suruga Bay to the Sea of Kumano Nada (Figs. 8.10 and 8.11).*

Bathymetric and c.s.p. surveys have been carried out by the Hydrographic Department in the Suruga Bay and by SATO and HOSHINO (1962), NAGANO *et al.* (1977) and OKUDA *et al.* (1976) in the Sea of Enshu Nada.

The Suruga Bay is the northeastern extension of the Nankai Trough. Its cross section has a V-shape, with a narrow flat bottomed floor which becomes stepwise deeper outwards from the margin. At Senoumi, in the bay entrance, samples of shale and sandstone of Miocene age were obtained. On the bay floor, thin mud beds cover the underlying thick sand-gravel beds. The latter contain pebbles of unaltered volcanic rocks, probably derived by submarine sliding from Quaternary volcanoes on the Izu Peninsula. (SATO, 1962).

The shelf edge is at -100 m depth off the Kii Peninsula, but becomes gradually deeper northeastwards, to southern offshore from Cape Daio $(-140$ to 150 m) and the Ise Bay $(-170$ m) to offshore from the Atsumi Peninsula $(-250$ to -290 m). Sediments on the shelf become thicker outwards, suggesting regional subsidence.

Four acoustic layers are classified by a c.s.p. survey: (1) the uppermost, (2) the reverberant, (3) the transparent and (4) the basement. They are correlated respectively with Quaternary , Pleistocene, Upper Miocene and pre-Lower Miocene (IWABUCHI *et al.*, 1976; OKUDA *et al.*, 1976).

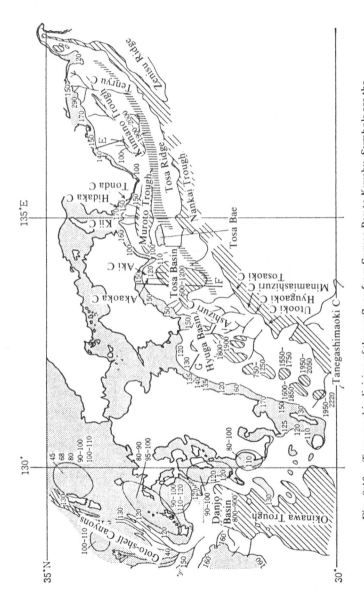

Fig. 8.10.   Topographic division of the sea floor from Suruga Bay to Kyushu. Symbols are the same as those in Fig. 8.4.

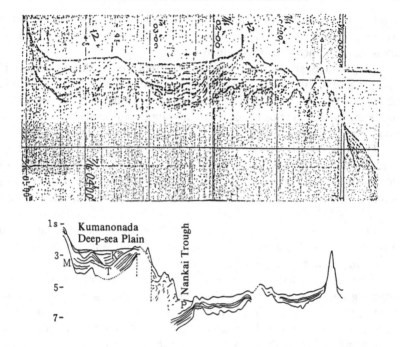

Fig. 8.11. Sonic survey record and cross-section of the Sea of Kumano Nada, along path E in Fig. 8.10. For key to symbols K, M, P and T, see text.

The deep-sea plain of the Sea of Kumano Nada is occupied by a turbidite-filled basin, which is divided into two parts by an E-W-trending anticline of the transparent acoustic layer. The reverberant layer, in the inner part, is traced northwards to the slope (Fig. 8.11). The outer margin of the plain is limited by the Tosa Ridge, composed of the seismically transparent layer and the basement.

vi) *Margin from Shikoku to Kyushu (Figs. 8.10 and 8.11)*

The shelf offshore from the Kii Peninsula to Shikoku is mostly 10 to 20 km in width, being 29 km in the area off Tosa. The shelf break mainly lies at -120 to -160 m, but becomes shallower near Cape Shionomisaki and Cape Muroto. The shelf-width and break depth to the east of offshore Kyushu are, respectively, 15 to 20 km and -120 to-170 m.

To the south of the Kii Strait, the acoustic basement is covered with layer T (Early to Middle Miocene), and the latter is overlain successively by Layer K1 (Middle Miocene to Early Pliocene), K2 (Middle to Late Pliocene), K3 (Early to Middle Pleistocene) and finally by P (Middle Pleistocene). The distribution of Layer K1 is limited in the Muroto Trough, K2 and K3 are more wide-spread and extend to the shelf and slope of the Kii Strait area.

Layer P comprises the turbidite fill of the northwestern part of the Muroto Trough. Tosa Bae (bank) is a part of the Tosa Ridge, and there Layers K2 and K3 overlie the acoustic basement.

The Tosa Basin forms a deep plain, −600 to −1,200 m deep, where Layers T and K form a syncline which is covered by Layer P. On the slope, to the east of the Ashizuri Peninsula, an unconformity is found at −750 m, suggesting basement subsidence with subsequent sedimentation on the slope. (SATO and SAKURAI, 1975).

The coastline of eastern Kyushu changes northwards; it is elevated in the Miyazaki Plain area and submerged in the Saeki and Usuki Bays. The shelf break also becomes deeper further northwards from −120 m to −140 m. The unconformity observed beneath the shelf is indistinct in the elevated coast region. (Fig. 8.12).

The continental slope to the east off Kyushu is underlain by flexured Miocene and overlying Pliocene beds. The largest submarine terrace on the slope is the Hyuga Terrace, which is a turbidite-filled basin.

From Suruga Bay to eastern Kyushu, the continental margin has the following common features: (1) flexured Miocene strata covered by Plio-Pleistocene sediments, (2) an unconformity particularly observed in bay areas, and (3) turbidite-filled deep-sea plains extending seawards. These features may have resulted from crustal warping movements since Neogene time (OKUDA et al., 1979).

vii)   *Western margin offshore Kyushu (Fig. 8.10)*

Studies along this area of the margin have been made by NAGANO *et al.*

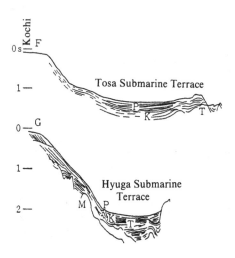

Fig. 8.12.   Cross-section along paths F and G in Fig. 8.10. For key to symbols K, M, P and T, see text.

(1976), SAKURAI and NAGANO (1976), KATSURA and NAGANO (1976), MIZUNO *et al.* (1971), TECHNICAL GROUP OF MARINE GEOLOGY SURVEY (1970, 1974), and GEOLOGICAL SURVEY OF JAPAN (1975a).

The shelf width is 30 km and reaches 40 km in the sea around the Goto Islands. The shelf break lies at -110 to -125 m to the south of Kagoshima and at -110 to -120 m at western off-shore Kyushu. A few N-E-stretching ridges are characteristic of the western Kyushu shelf.

The Danjo Basin, which is the northernmost part of the Okinawa Trough, lies between the shelf of western Kyushu and that of the East China Sea. The Goto Submarine Canyon is developed between the Goto and Danjo Islands. The shallower part of the canyon branches arborescently into the Goto Shelf Channels.

The acoustic basement (Layer E) of this region is considered to be older than Middle Neogene in age. It is successively covered by Layers D (Late to latest Miocene), C (Pliocene) and B (Lower Pleistocene).

The Tsushima and Iki Islands, Goto Islands, Nagasaki Peninsula, Koshiki Islands and Danjo Islands commonly show a NNE-SSW to NE-SW trend, while the Goto Submarine Canyons and the Goto Shelf Channels lie in a NW-SE direction.

viii) *Nansei-shoto marginal area, including the East China Sea (Fig. 8.13)*

The GEOLOGICAL SURVEY OF JAPAN published a submarine geological map of this area (1979b), and HAMAMOTO *et al.* (1979) made a study, particularly in the Miyako-Yaeyama Islands area.

The topography of this area is classified, from northwest to southeast,

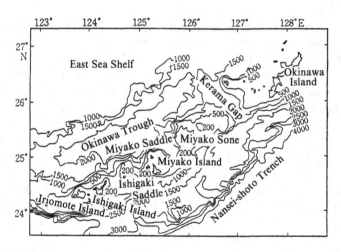

Fig. 8.13. Submarine topography around the Miyako-aeyama Islands (after HAMAMOTO *et al.*, 1979).

into: The East China Sea Shelf, Okinawa Trough, the volcanic arc, Nansei-shoto Islands and Ryukyu Trench.

The Ryukyu Ridge comprising the volcanic arc and Nansei-shoto Islands, is divided into a few sectors by saddles and gaps. They are, from south to north: the Ishigaki, Tarama and Miyako Saddles and the Kerama Gap. The last is very deep, −1940 m, but to the south it continues to a submarine plateau, on which a few banks occur, for example, Miyako Sone Bank.

The volcanic arc is emerged in the northern, but submerged in the southern part, the latter being continued to the West Ryukyu Submarine Hills.

The acoustic stratigraphy is classified into Layer A (Pleistocene, reverberant), B (Pliocene, transparent) and C (Miocene to Lower Pliocene, opaque).

The Okinawa Trough forms a basin shallower than −2,500 m depth and filled with Plio-Pleistocene strata thicker than 5,000 m, forming a delta-like monoclinal structure. The East China Sea is a vast continental shelf covered by Neogene strata whose thickness is greater than 2,000 m. Along the margin of the East China Sea , a ridge extends from the Goto to the Senkaku Islands. The shelf break is fairly deep, being −220 to −250 m deep near the Senkaku Islands, but submarine terraces of −135 to 145 m and −180 m are also reported on the shelf.

8.2   Ocean Floor Around the Japanese Islands

8.2.1   Outline (Fig. 8.1 & Fig. 8.14)

The Western Pacific is a boundary area between the Asian Continental mass and the Pacific Ocean. The Pacific plate is being subducted under the Asian Continent, resulting in the formation of island arcs and associated marginal seas, which are characteristic of subduction zones. The Western Pacific is divided into the Northwestern Pacific Basin and the Mariana Basin. The boundary of the two is marked by the Mid-Pacific Seamounts.

The western marginal trenches of the Pacific plate are divided into groups, which are, from north to south: the Kurile-Kamchatka, Japan, Izu-Ogasawara and Mariana Trenches. The associated island arcs are, respectively, the Kurile Islands and eastern Hokkaido, Northeast Japan, the Izu-Ogasawara and Mariana Islands. These arcs embay marginal seas such as the Okhotsk Sea, the Sea of Japan and the Philippine Sea. The Philippine Sea is divided by the Kyushu-Palau Ridge into the Shikoku-Parece Vela Basin to the east, and the Philippine Basin to the west.

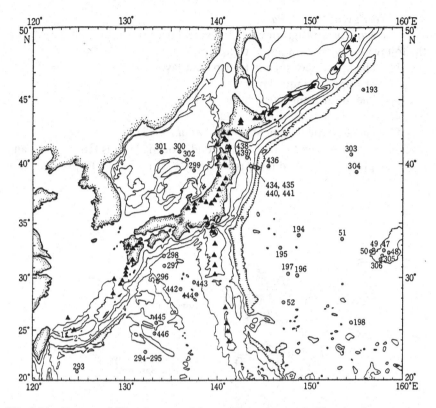

Fig. 8.14.   Deep-sea drilling sites around Japan. Circle with dot: Drilling site; figures indicate the site numbers. Solid triangle: Quaternary volcano. Numbers on contour lines: depth in km. Data in the following table.

| site | Leg | Leg Period | Reported in | |
			Geotimes	Init. Rept.
44~60	6	1969 VI–VIII	1969 X	1971 II
183~193	19	1971 VII–IX	1971 XI	1973 VI
194~202	20	1971 IX–XI	1972 IV	1973 XI
290~302	31	1973 VI–VIII	1973 X	1975 V
303~313	32	1973 VIII–X	1973 XII	1975 VIII
434~437	56	1977 IX–X	1978 III	
438~441	57	1977 X–XI	1978 IV	
442~446	58	1977 XII–I	1978 V	

8.2.2 Northwest Pacific Basin

The acoustic stratigraphy of this typical oceanic basin is classified into
the following layers from upper to lower:

1) The upper acoustically transparent layer,
2) The upper opaque layer,
3) The lower transparent layer,
4) The lower opaque layer, and
5) The acoustic basement (EWING *et al.*, 1968).

The boundary between the first and second acoustic layer is Horizon 'A', and
the basal plane of the third layer is Horizon 'B' (Fig. 8.15). Deep-sea drilling

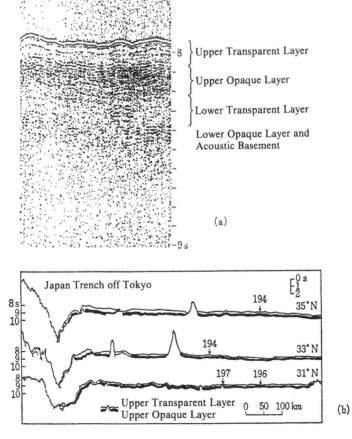

Fig. 8.15.   Sonic survey record in Northwest Pacific (after Init. Rept. DSDP, 20). (a) Record
at Site 195, (b) E-W cross sections along three latitudes.

has enabled the lithology of the acoustic strata to be defined as follows: (Fig. 8.16).

1)   Upper transparent layer: Clayey silt and brown clay rich in volcanic materials. Less than 250 m in thickness.

2)   Upper opaque layer: Chert and chalk of Middle to Late Cretaceous. 60 to 100 m in thickness. Reflectivity varies from place to place.

3)   Lower transparent layer: Pelagic clay. 10 to 50 m in thickness.

4)   Lower opaque layer: Chert and chalk. In some places reaches 150 m in thickness.

5)   Basement: Oceanic basalt.

Sediments on the Schatsky Rise are highly suitable for study of planktonic microfossils, for here the sediments are not co-mingled with terrestrial materials and the fossils have been well preserved because the depth is shallower than the CCD (Fig. 8.17). As a result, the following correlation has been established.

1)   Upper transparent layer: Nanno-fossil ooze of late Cretaceous to Recent age. In the parts deeper than CCD, it consists of siliceous ooze and brown clay with volcanic materials. Non-depositional and unconformable gaps are found in the Middle Miocene. The layer is 200 to 300 m in thickness.

2)   Upper opaque layer: Middle Cretaceous chert and chalk. About 500 m in thickness.

3)   Lower transparent layer: Early Cretaceous chalk and pelagic clay.

4)   Lower opaque layer: Late Jurassic to Early Cretaceous chert.

5)   Basement.

The lithology of the upper transparen layer, traced over the whole Northern Pacific region, is not uniform, varying from chert through tuff and chalk to radiolarian clay, and its age is also variable, changing from late Cretaceous to Oligocene.

Studies of magnetic anomalies in this region were analysed by HILDE et al. (1976) and the magnetic stripes were numbered as CL and M1 to M26 (Fig. 8.18). The stripes run parallel in a NE-SW direction and are disrupted by NW-SE trending oceanic fracture zones. The age of sediments covering the ocean-floor basalt, over a few magnetic stripes have been tabulated in Table 8.1, and the magnetic directions and paleo-latitudes measured for seamounts are tabulated in Table 8.2.

There are a number of seamounts in the northwest Pacific, particularly in its southern part. These are frequently guyots, with summits commonly at −1,500 to −2,000 m depth. Shallow-marine fossils, tropical molluscs, corals and echinoderms, obtained from the truncated tops, are always of Cretaceous age, and the radiometric ages of volcanic rocks collected with the fossils indicate 90 to 100 Ma (HEEZEN et al., 1973).

Peculiar magnetic anomalies are observed in the seamounts. An analysis

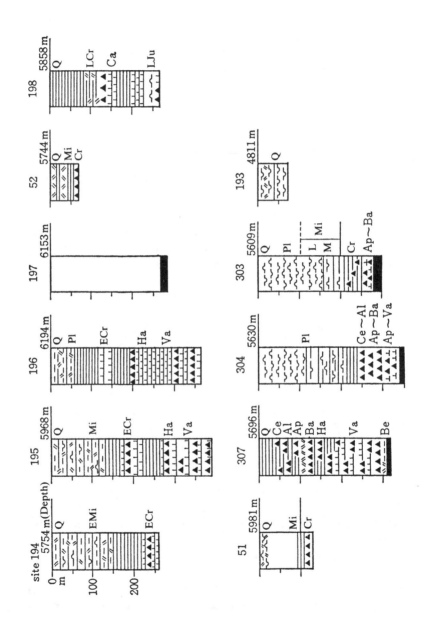

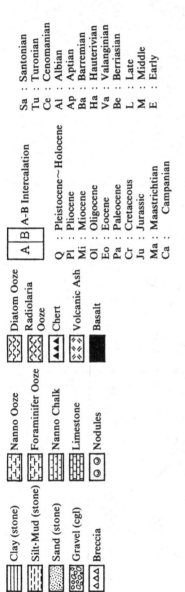

Fig. 8.16. Columnar sections obtained by deep-sea drilling in Northwest Pacific. The site locations are shown in Fig. 8.14.

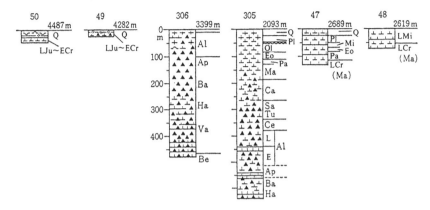

Fig. 8.17.   Columnar sections in the Schatsky rise region. Symbols are the same as those given in Fig. 8.16. The site locations are given in Fig. 8.14.

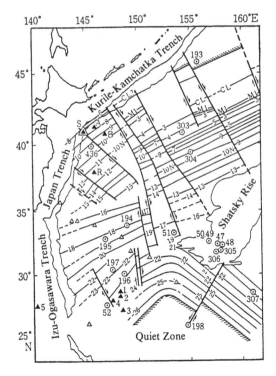

Fig. 8.18.   Magnetic anomalies in Northwest Pacific (after HILDE *et al.*, 1976). Circle with dot: Deep-sea drilling site. Solid triangle: Seamounts, on which magnetic directions were measured (see Table 8.2). Open triangle: Guyots, from whose summits fossils were collected (after HEEZEN *et al.*, 1973; TSUCHI and KURODA, 1973; SHIBATA, 1979).

Table 8.1.   Magnetic anomaly numbers and fossil ages of the Deep-sea drilling sites.

Magnetic Anomaly	Fossil Age	DSDP site	Paleo-latitude
M4	Hauterivian, 124±12 Ma	303	6°S
M9	Valanginian, 127±9 Ma	304	11°S
M21	Berriasian, 136±12 Ma	307	6°N

Table 8.2.   Magnetic directions estimated from magnetic anomalies observed at seamounts. The locations are given in Fig. 8.18 (after UYEDA and RICHARDS, 1966; VACQUIER and UYEDA, 1967).

	Latitude	Longitude	Intensity ($\times 10^{-2}$ emu/cc)	Declination	Inclination	Magnetic Pole °N	Magnetic Pole °E	Paleo-latitude
A	41°16′N	145°58′E	1.56	−7°	2°	49	−23	1°N
S	40°55′N	144°54′E	1.04	341°	22°	56	−1	12°N
B	40°38′N	146°51′E	0.392	353°	−4°	49	−22	2°S
R	38°00′N	145°58′E	0.603	344°	3°	50	−8	1°N
1	28°48′N	148°21′E	0.183	334°	9°	55	19	4°N
2	28°22′N	148°14′E	0.419	28°	5°	53	−82	2°N
3	27°03′N	148°39′E	0.643	16°	−13°	53	−58	7°S
4	27°57′N	147°34′E	0.299	11°	−1°	60	−54	1°S
5	27°41′N	140°24′E	0.315	5°	39°	82	−78	22°N

of the anomalies gives the paleo-latitudes of the seamounts, which suggest that they were located in low-latitudes areas or even in the Southern Hemisphere at the time of their formation. The data are consistent with the fossil evidence. The northward movement of the seamounts is in agreement with the direction of Pacific plate motion, inferred from a study of the Emperor and Hawaiian chains.

8.2.3   Vicinity of Trenches
i)   *Japan Trench and environs*

Topographic swells, 200 to 500 km in width, and about 500 m in relative height, run on the oceanward side of the Kamchatka-Kurile, Japan and Izu-Bonin Trenches, but they are not observed to be associated with the Mariana Trench.

The acoustic stratigraphy of the outer topographic swell is similar to that of the Northwest Pacific Ocean. The upper acoustically transparent layer consists of diatomaceous ooze, and the upper opaque layer is composed of Cretaceous chert. The boundary 'A' between the two indicates non- or sparse sedimentation in Paleogene time.

On the outer trench slope, grabens resulting from normal faulting run parallel to the trench axes (IWABUCHI, 1968a). They are about 2 km in width and −200 to −400 m in depth.

The acoustic basement is further subducted underneath the island arcs (Fig. 8.19). This is in agreement with the magnetic anomaly pattern that can be traced to the island arc area beyond the trench (HILDE et al., 1976).

The flat bottomed floor of the Japan Trench is filled with turbidites and becomes deeper, stepwise to the south from its junction with the Kamchatka-Kurile Trench. In the area to the south of Joban, sediments are almost absent and the trench here has a V-shaped section.

At the junction of the Kamchatka-Kurile and Japan Trenches, a few seamounts occur, such as the Erimo Seamount, the Daiichi Kashima Seamount and further south, the Ogasawara Plateau. Their summits are commonly 1,500 to 2,000 m in depth. The flat top of the Erimo Seamount dips northwestwards, being parallel to the plane of the subducting plate. Cretaceous shallow-marine fossil shells, e.g. *Nerinea (Plesioptygmatis) ryofuae*, and algae were dredged from the Erimo Seamount. On the other hand, the older sediments dredged from the Daiichi Kashima Seamount are considered to be of Barremian to Cenomanian age, based on fossils such as *Astroprina* (Mesozoic), *Orbitolina* (Barremian to Cenomanian) and *Cuneolina* (Albian to Miocene) (TOKAI UNIVERSITY SURVEY GROUP, 1976). On the Yabe Seamount to the west of the Bonin Plateau *Nerinea pauxilla*, *Pachyrismella* ? sp. occurs abundantly together with other fossils of Middle Cretaceous age. (SHIBA, 1979).

The landward slope has a steep and ragged surface: the ridge and trough zone as defined by IWABUCHI et al. (1979). It is caused by an imbricate structure, consisting of numerous wedge-shaped thrust sheets dipping landwards. The structure is called an accretionary prism.

The landward slope of the Japan Trench off Sanriku is underlain by the thrusted sheets, mentioned above, but on the slope a few basinal structures occur consisting of slump deposits (Fig. 8.20). Drilling led to the recovery of thick, 637.5 m, core of sediments younger than Late Miocene age. A study of diatom remains indicated that one horizon recurs many times in the core. Furthermore, sediments are fairly indurated, and are strongly sheared and cut by numerous faults.

Drilling on the landward slope, off Hachinohe, reached the acoustic basement at −1,100 m depth (Fig. 8.19). The basement consists of dacite boulder conglomerate of Late Oligocene age, unconformably covering black siliceous claystone of Late Cretaceous age. The conglomerate is successively overlain by Late Oligocene massive sandstones, yielding abundant fossils,

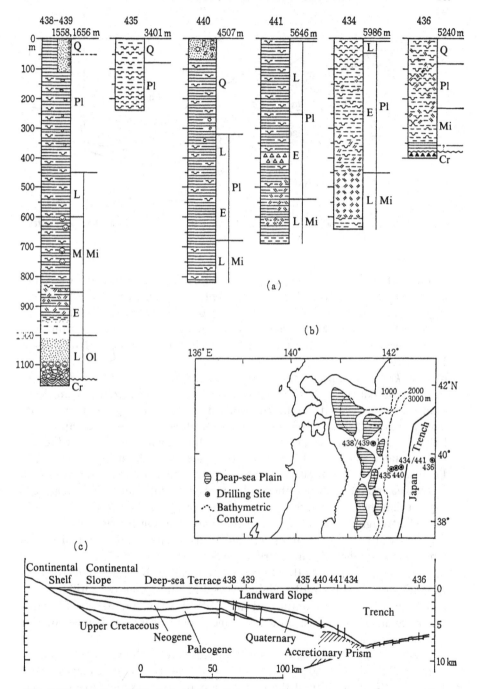

Fig. 8.19. Deep-sea drilling in Japan Trench. Symbols are the same as those given in Fig. 8.16
(b: modified after IWABUCHI, 1969; c: after VON HUENE et al., 1978).

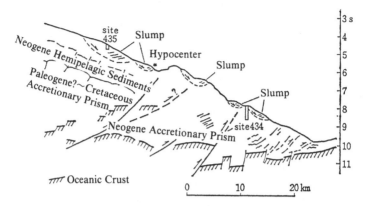

Fig. 8.20.   Sonic profile taken on the landward slope of the Japan Trench, at about 40 N (after
LANGSETH *et al.*, 1978).

and by turbidites and claystones of Early Miocene age, and further overlain
by younger diatomaceous clay. The sedimentary sequence above the Miocene
suggests a temporary change in the rate of deposition; that is, the rate
decreased in the Late Miocene, with practically no sedimentation during the
Late Pliocene and the rate increased again in the Pleistocene (NIITSUMA,
1978, 1979).

ii)   *Nankai Trough and Nansei-shoto Trench (Figs. 8.10, 8.13 and 8.21)*
      The Nankai Trough is not as deep as a trench (*sensu stricto*), being less
than −6,000 m, even if the thickness of sediments overlying the floor of the
trough are disregarded. However, the seismic plane indicates subduction at a
low angle of less than 20°, under Southwest Japan, which is considered to be
the associated island arc. The depth of the plane lies at about 50 km in the Kii
Peninsula and at about 80 km in Shikoku. On the other hand, the volcanic·
activity has not been clearly observed along this seismic plane. These features
suggest that the Nankai Trough is probably younger than 5 Ma.

      The trough sediments are distinctly deformed. On the floor of Suruga
Bay, eastwards tilting is observed, and in the region to the west of Cape
Omaezaki a number of folded structures occur in the sediments.

      On the lower part of the slope is an area called the "ridge and trough"
zone of complex topography similar to that mentioned above. Drilling was
cored out on the lower continental slope, off Shikoku, and 611 m of highly
indurated and sheared unstratified siltstone were recovered, all being younger
than Pleistocene.

      The southern part of the Nansei-shoto Trench shows typical trench
topography, which becomes indistinct to the northeast of Kikaigashima.
However, the continuity of the volcanic arc and the seismic plane suggests

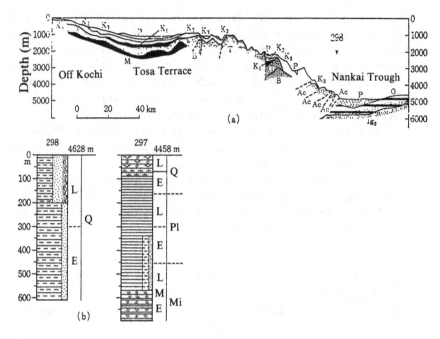

Fig. 8.21.   Cross and columnar sections in the area off Kochi. The surveying path is shown in Fig. 8.23. (a) Cross section (after GEOL. SURV. JAPAN, 1977c). P and K_3: Recent-Pleistocene, K_2: Upper Pliocene, K_1: Lower Pliocene-Upper Miocene, T: Middle Miocene, M: Middle Miocene-Oligocene, B: Acoustic basement, Ac: Recent-Pliocene, O: Recent-Miocene, Ig_2: Igneous rocks. (b) Columnar-sections at Sites 297 and 198. Symbols are the same as those in Fig. 8.16.

that the trench is connected with the Nankai Trough to the east of offshore Kyushu.

The Yaeyama Basin lies offshore from the Yaeyama Islands. It is filled with more than 4,000 m of sediments, the lower part of which may be correlated with the Early Miocene Yaeyama Formation of Iriomote-Jima, while the upper part is turbiditic and is correlated with the Late Miocene to Pliocene Shimajiri Formation. On the other hand, the Shimajiri Basin to the northeast is underlain by the Shimajiri Formation and Ryukyu Limestone (Pleistocene) (AIBA and SEKIYA, 1979).

8.2.4   Sea of Japan

The Sea of Japan is roughly divided into a northern and a southern part. In the northern Sea of Japan, the Japan Basin has a flat floor at about 3,500 m depth, and an oceanic crustal thickness of about 8,000 m. The southern Sea of Japan has an uneven floor formed of continental crust and has such features as the Yamato Ridge, Yamato Basin and the Korean Plateau.

Various rock pebbles such as granite, phyllite and chert have been obtained from the Yamato Ridge, and those of rhyolite and dacite correlated with the Nohi Rhyolite of central Japan, from the Oki-Yamato Ridge. Gneissose granite pebbles obtained from the Yamato Ridge have been dated as 197 Ma and 220 Ma by the K–Ar method (IWABUCHI, 1968a). Similarly, granite, acidic welded tuff and intermediate to mafic volcanic breccia were collected from various rises in the Sea of Japan, and Late Pliocene (2.0 to 3.0 Ma) siltstone from eastern Yamato Tai (GEOLOGICAL SURVEY OF JAPAN, 1979). Furthermore, alkalic volcanic rocks were reported from Yamato and Shun-yo Tais (IWABUCHI, 1968a).

Deep-sea drilling carried out in the Toyama Deep-sea Channel (Site 299), Sea of Japan (Site 301) and the northeastern part of Kita-Yamato Tai (Site 302) gave the following correlation between the acoustic and lithological stratigraphy (KOIZUMI, 1977) (Fig. 8.22).

1)   Upper opaque layer: Plio-Pleistocene turbidite and diatomaceous clay. Thicker than 500 m in the Japan and Yamato Basins.

2)   Upper transparent layer: Late Miocene to Early Pliocene diatomaceous ooze. About 200 m thick in northeastern Yamato Tai.

3)   Lower opaque layer: Miocene zeolite-bearing clay-claystone, greenish-gray silty clay and greenish tuff.

4)   Lower transparent layer: Diatomaceous ooze.

Between Sakhalin and Sikhotealin there is the Tartary Trough, whose central part is underlain by the acoustically opaque layer reaching a thickness greater than 2,000 m.

In the western part of the Japan Basin to the east of the Korean Peninsula is the Korean Plateau (or rise). It consists of two swells extending

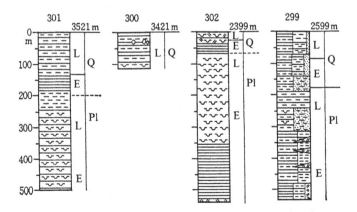

Fig. 8.22.   Columnar-sections at the drilling sites in the Sea of Japan. The locations are given in Fig. 8.14, and symbols are the same as those given in Fig. 8.16.

in an ENE direction. Pebbles of granite and gneiss, probably of Late Jurassic to Early Cretaceous age were obtained from the plateau (GEOLOGICAL SURVEY OF JAPAN, 1978a).

8.2.5   Philippine Sea (Fig. 8.23)
i)   Daito Ridges

The Daito Ridges comprises the Amami Plateau, Daito Ridge and Okidaito Ridge. They are 2,000 to 3,000 m in depth, and on their summits and slopes are found various specimens, among which *Nummulites* limestone are to be mentioned (Fig. 8.23). The *Nummulite*-bearing beds overlie the acoustic basement and can be traced over wide areas of the Philippine Sea Basin. Radiometric ages of igneous rocks obtained from the Amami Plateau are 82 to 85 Ma (MIZUNO *et al.*, 1976; SHIKI *et al.*, 1975).

Deep-sea drilling was performed on the Daito Ridge (Site 445) and in the Daito Basin (Site 446). The older sediments recovered are mudstone, siltstone, sandstone and conglomerate, the latter two formed by redeposition due to turbidity current which carried *Nummulites* into the sediment. On the other hand, the younger sediments are pelagic nannoplanktonic ooze, chalk

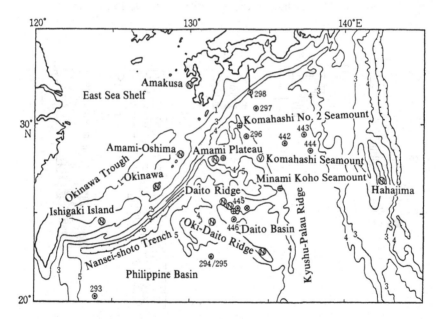

Fig. 8.23.   Localities of *Nummulites*, igneous rocks and metamorphic rocks in Philippine Sea. Circle with N: *Nummulites*, Circle with cross: Plutonic rocks, Circle with v: Volcanic rocks, Circle with diagonal cross: Metamorphic rocks, Circle with dot: Deep-sea drilling site, Thick line: Line of cross-section given in Fig. 8.12, Numeral on contour: Depth in km.

and limestone on the ridge and dark-brownish clay and tuff in the basin. This contrast indicates that the CCD occurred at the depth between the ridge and the basin.

Early Eocene mudstone beds in the Daito Basin were intruded by some dozen sills of tholeiitic basalts (Fig. 8.24).

ii) *Shikoku Basin*

The depth of the Shikoku Basin is about 4,600 m, though the eastern part is slightly shallower than the western part. The deposits on the floor are relatively thinner in the N-S-trending axial part, but are thicker on both sides. The spreading axis is inferred from the fairly well-studied magnetic anomalies. The Deep-Sea Drill Program drilled one site (Site 442) on the western and at two sites (Sites 443 and 444) on the eastern side of the axis. Middle Miocene to Recent hemipelagic clay intercalated with nanoplanktonic ooze were recovered (Fig. 8.25).

Sediments older than Middle Miocene are interlayered with lavas and sills of abyssal tholeiite. At Site 442 sediments of 15–17 Ma are underlain by massive basalt, under which 18–21 Ma sediments are interbedded with pillowy basalt. On the other hand, at Sites 443 and 444, 15 Ma sediments lie just over the massive basalt. It is suggested, therefore, that widespread basaltic volcanism took place in the Shikoku Basin ca. 15 m.y. ago.

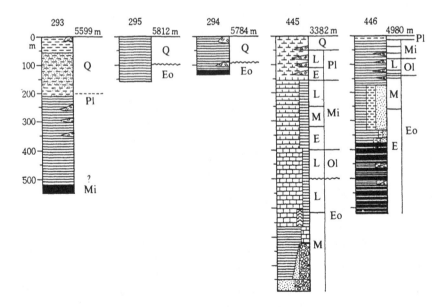

Fig. 8.24. Columnar-section at the drilling sites in Philippine Sea. Symbols are the same as those given in Fig. 8.16.

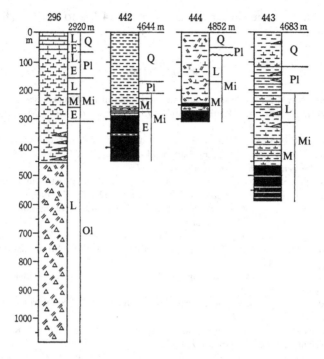

Fig. 8.25. Columnar sections at the drilling sites in the Shikoku Basin. Symbols are the same as those given in Fig. 8.16.

iii) *Kyushu-Palau Ridge*

This ridge is mostly underlain by acoustic basement, while a veneer of sediments cover the crest and part of the slope. The sediments recovered at Site 296 largely consist of Oligocene pyroclastic materials, overlain by Early Miocene to Recent pelagic nannoplankton ooze. Similar volcanic materials of Oligo-Miocene age were also obtained at the Komabashi Seamount, and may be compared with those on the Guam and Saipan Islands. The basement rocks comprise quartz diorite, tonalite, trondhjemite and granodiorite of 37 to 48 Ma.

8.2.6 *Kurile Basin*

Being similar to the Japan Basin, the Kurile Basin, 3,200 to 3,300 m deep, is underlain by oceanic crust. Sediments thicker than 3,000 m cover the basement rocks and show the features characteristic of an abyssal plain.

The continental shelf of the Okhotsk Sea to the north of the Kurile Basin is underlain by sediments thinner than 1,000 m. From the exposed basement rock samples were obtained: granodiorite, porphyrite and biotite schist, whose radiometric ages range from 73–94 Ma to 209 Ma (BURK and GNIBIDENKO, 1977).

REFERENCES

AIBA, J. and SEKIYA, E., Distribution and characteristics of the Neogene sedimentary basins around the Nansei-shoto (Ryukyu islands), *J. Japan. Assoc. Petrol. Tech.*, **44**, 329–340, 1979.

BURK, C. A. and GNIBIDENKO, H. S., The structure and age of acoustic basement in the Okhotsk Sea. Island Arcs, Deep Sea Trenches and Back-arc Basins (M. Talwani and W. C. Pitman, III, eds.), *Amer. Geophys. Union*, 451–461, 1977.

EWING, J., EWING, M., AITKEN, T. and LUDWIG, W. J., North Pacific sediment layers measured by seismic profiling, *Amer. Geophys. Union Monogr.*, **12**, 147–173, 1968.

GEOLOGICAL SURVEY OF JAPAN, Goto-nada and Tsushima Strait investigation, northwestern Kyushu, *Cruise Report No. 2*, 1975a.

GEOLOGICAL SURVEY OF JAPAN, Sagami-nada Sea investigation, *Cruise Report No. 3*, 1975b.

GEOLOGICAL SURVEY OF JAPAN, Geological Map of the South of Kii Strait 1:200,000, Marine Geology Map Series 5, 1977a.

GEOLOGICAL SURVEY OF JAPAN, Geological Map around Ryukyu Arc 1:1,000,000, Marine Geology Map Series 7, 1977b.

GEOLOGICAL SURVEY OF JAPAN, Geological Map off Outer Zone of Southwest Japan, Marine Geology Map Series 8, 1977c.

GEOLOGICAL SURVEY OF JAPAN, Geological investigation of Japan and southern Kurile Trenches and slope area, *Cruise Report No. 7*, 1977d.

GEOLOGICAL SURVEY OF JAPAN, Geological Map of the Japan and Kurile Trenches and the Adjacent Areas 1:1,000,000, Marine Geology Map Series 11, 1978a.

GEOLOGICAL SURVEY OF JAPAN, Geological investigations in the northern margin of the Okinawa trough and the western margin of the Japan Sea, *Cruise Report No. 10*, 1978b.

GEOLOGICAL SURVEY OF JAPAN, Geological investigation of the Okhotsk and Japan Seas off Hokkaido, *Cruise Report No. 11*, 1978c.

GEOLOGICAL SURVEY OF JAPAN, Geological investigation of the Japan Sea, *Cruise Report No. 13*, 1979.

GROUP OF MARINE GEOLOGICAL SURVEY TECHNIQUES, Marine geological survey in environs of Koshiki island off western Kyushu, *Chishitsu News*, **186**, 26–36, **188**, 12–20 and **190**, 1–30, 1970.

GROUP OF MARINE GEOLOGICAL SURVEY TECHNIQUES, The outline of marine geological survey researches in environs of Goto and Tsushima islands, *Chishitsu News*, **233**, 1–18, 1974.

HAMAMOTO, F., SAKURAI, M., and NAGANO, M., Submarine geology off the Miyako and Yaeyama islands, *Rept. Hydrogr. Res.*, **14**, 1–38, 1979.

HEEZEN, B. C., MATTHEWS, J. L., CATALANO, R., NATLAND, J., COOGAN, A., THARP, M., and RAWSON, M., Western Pacific guyots, *Init. Rept. DSDP*, **20**, 653–723, 1973.

HILDE, T. W. C., ISEZAKI, N., and WAGEMAN, J. M., Mesozoic sea-floor spreading in the north Pacific, *Amer. Geophys. Union Monogr.*, **19**, 205–226, 1976.

HONZA, E., Tectonic history of the Northeast Japan island arc since Neogene, *Marine Geology*, University of Tokyo Press, 137–154, 1976.

HOSHINO, M. and SATO, T., On the topography and bottom sediment of Kamogawa submarine canyon, Boso Peninsula, *Quatern. Res.*, **1**, 228–237, 1960.

HOTTA, H., The structure of sedimentary layer in the Japan Sea, *Bull. Hokkaido Univ.*, **18**, 111–131, 1967.

IWABUCHI, Y., Submarine geology of the southeastern part of the Japan Sea, *Contrib. Inst. Geol. Paleontol., Tohoku Univ.*, **66**, 1–76, 1968a.

IWABUCHI, Y., Topography of trenches east of the Japanese Islands, *J. Geol. Soc. Japan*, **74**, 37–46, 1986b.

IWABUCHI, Y., Topographie des fosses au Pacifique de nord-ouest. *La Mer*, **7**, 197–205, 1969.

IWABUCHI, Y., KATSURA, T., NAGANO, M., and SAKURAI, M., Submarine geology of Fossa Magna region, *Marine Sciences/Monthly*, **8**, 549–556, 1976.

KATSURA, T. and NAGANO, M., Geomorphology and tectonic movement of the sea floor, northwest off Kyushu, Japan, *J. Oceanogr. Soc. Japan*, **32**, 139–150, 1976.

KIMURA, M., Geological structure around South Kanto region, Japan, *Marine Geology*, University of Tokyo Press, 155–176, 1976.

KOIZUMI, I., Pelagic sediments in relation to the history of Japan Sea, *Kagaku (Science)*, **47**, 45–51, 1977.

KOMATSU, N., Offshore sedimentary basin, northeastern part of Honshu, Japan, *J. Japan Assoc. Petrol. Tech.*, **44**, 268–271, 1979.

LANGSETH, M. G., OKADA, H. *et al.*, Transects began near the Japan Trench, *Geotimes*, **23**, 3, 22–26, 1973.

MIZUNO, A. and GROUP OF MARINE GEOLOGICAL SURVEY TECHNIQUES, Geology off western Kyushu, Japan, *Geological Problems in the Sea around Kyushu*, 61–70, 1971.

MIZUNO, A., OKUDA, Y., and TAMAKI, K., Some problems on the geology of the Daito Ridges region and its origin, *Geological Studies of the Ryukyu Islands* (K. Kizaki, ed.), **1**, 177–198, 1976.

MOGI, A., On the depth of the continental shelf margin along the north-west coast of Honshu, *Hydrogr. Bull. Spec.*, No. 12, 54–57, 1953.

MOGI, A., The submarine topography in the eastern and western parts of Sagami Bay, *Hydrogr. Bull. Spec.*, No. 17, 115–127, 1955.

MOGI, A., Drowned topographies in the near shore bottom of Yufutsu Plain, Hokkaido, *Quatern. Res.*, **3**, 141–152, 1964.

MOGI, A. and IWABUCHI, Y., Submarine topography and sediments on the continental shelves along the coasts of Joban and Kashimanada, *Geogr. Rev. Japan*, **34**, 159–177, 1961.

MURAUCHI, S. and ASANUMA, T., Studies on Seismic profiler measurements off Boso-Jyoban district, northeast Japan, *Bull. Nat. Sci. Mus.*, **13**, 337–356, 1970.

NAGANO, M., AJIRO, T., and TOZAKI, T., Submarine geology of Ensyu-nada Sea, south of Honshu, *Rept. Hydrogr. Res.*, **12**, 1–33, 1977.

NAGANO, M., SAKURAI, M., KATSURA, T., NAKAMURA, H., KITAHARA, S., and ONODERA,K., Submarine geology off west coast of Kyushu, *Rept. Hydrogr. Res.*, **11**, 1–38, 1976.

NAGANO, M., SAKURAI, M., UCHIDA, M., IKEDA, K., TAGUCHI, H., and OMORI, T., Submarine geology off northeast coast of Hokkaido district, *Rept. Hydrogr. Res.*, **9**, 1–31, 1974.

NAKAJIMA, T., Submarine topography off the southern part of Sanriku district, *J. Geogr.*, **82**, 136–147, 1973.

NIITSUMA, N., Magnetic stratigraphy of the Japanese Neogene and the development of the island arc of Japan, *J. Phys. Earth*, **26**, 367–378, 1978.

NIITSUMA, N., Geotectonic development of Northeast Japan Arc, *Kagaku (Science)*, **49**, 36–43, 1979.

OKAMOTO, K. and HONZA, E., The "Pliocene" fossil molluscan assemblage including Amussiopecten collected by GH 77-2 cruise in the southwestern Japan Sea, *J. Geol. Soc. Japan*, **84**, 625–628, 1978.

OKUDA, Y., INOUE, E., ISHIHARA, T., KINOSHITA, Y., TAMAKI, K., JOSHIMA, M., and ISHIBASHI, K., Submarine geology of the Nankai Trough and its peripheral area, *Marine Sciences/Monthly*, **8**, 192–200, 1976.

OKUDA, Y., KUMAGAI, M., and TAMAKI, K., Tectonic development of the continental slope and its peripheral area off Southwest Japan in relation to sedimentary sequences in sedimentary basins, *J. Japan. Assoc. Petrol. Tech.*, **44**, 279–290, 1979.

RESEARCH GROUP OF DAIICHI-KASHIMA SEAMOUNT, Topography and geology of Daiichi-Kashima seamount, off Inubo Cape, southeastern Honshu, Japan, *Chikyu Kagaku (Earth Science)*, **30**, 222–240, 1976.

SAKURAI, M. and NAGANO, M., Submarine topography and geological structure west of Kyushu, Japan, *J. Geogr.*, **85**, 329–341, 1976.

SAKURAI, M., NAGANO, M., NAGAI, T., KATSURA, T., TOZAWA, M. and IKEDA, K., Submarine geology off the south coast of Hokkaido, *Rept. Hydrogr. Res.*, **10**, 1–37, 1975.

SATO, T., Submarine topography in the environs of the Kushiro submarine canyon, *J. Geol. Soc. Japan*, **68**, 563–572, 1962a.

SATO, T., Sand and gravel bed cored from the bottom of the Suruga bay, *J. Geol. Soc. Japan*, **68**, 609–617, 1962b.

SATO, T., *Submarine geology of the continental shelves around the Japanese Islands*, Kyoritsu Shuppan Co., 193 p., 1970.

SATO, T., Sea bottom survey in westward of Northeast Japan, *J. Geogr.*, **80**, 285–301, 1971.

SATO, T., Several considerations on the deep sea plains, *Marine Sciences/Monthly*, **5**, 1973.

SATO, T., Sea level change deduced from submarine topography and geology around Japan, *Earth Monthly*, **1**, 392–397, 1979.

SATO, T. and HOSHINO, M., Submarine geology of the Enshu-nada Sea, southern Honshu, *J. Geol. Soc. Japan*, **68**, 313–328, 1962.

SATO, T. and SAKURAI, M., Submarine geological structure in the continental slope off southern Shikoku, *Geological Problems in the Philippine Sea*, 5–10, 1975.

SATO, T., SAKURAI, M., TAGUCHI, H., NAGANO, M., UCHIDA, M., and OMORI, T., Submarine geology of the continental borderland west off Hokkaido, *Rept. Hydrogr. Res.*, **8**, 1–49, 1973.

SHIBA, M., Geological history of the Yabe guyot to the east o the Ogasawara islands, *J. Geol. Soc. Japan*, **85**, 209–220, 1979.

SHIKI, T., TOKUOKA, T., AOKI, H., MISAWA, Y., KONDA, K., and NISHIDA, S., GDP-cruises in the Philippine Sea, especially on the results of dredges in GDP-8 and -11, *Geological Problems in the Philippine Sea*, 67–74, 1975.

TAGUCHI, H., NAGANO, M., SATO, T., SAKURAI, M., and UCHIDA, M., Structural development of the area adjacent to the Hakusan Se, Japan Sea, *J. Geol. Soc. Japan*, **79**, 287–298, 1973.

TSUCHI, R. and KURODA, N., Erimo (Sysoev) seamount and its relation to the tectonic history of Pacific Ocean Basin, The Western Pacific: Island Arcs, *Marginal Seas, Geochemistry* (P. J. Coleman ed.), Univ. West Australia Press, 57–64, 1973.

UTASHIRO, S. and SATO, T., A sub-bottom in the environs of the 1968 Earthquake epicenter, *Report on the Tokachi-oki Earthquake 1968*, 307–319, 1969.

UYEDA, S. and RICHARDS, M., Magnetization of four Pacific seamounts near the Japanese Islands, *Bull. Earthq. Res. Inst.*, **44**, 179–213, 1966.

VACQUIER, V. and UYEDA, S., Paleomagnetism of nine seamounts in the Western Pacific and of three volcanoes in Japan, *Bull. Earthq. Res. Inst.*, **45**, 815–848, 1967.

VON HUENE, R. and others, On Leg 56, Japan Trench transected, *Geotimes*, April, 1978, 16–21.

YAMAMOTO, H., The geologic structure and the sedimentary basins of northern part of the Hokkaido island, *J. Japan. Assoc. Petrol. Tech.*, **44**, 260–267, 1979.

CHAPTER 9

Chapter 9

ORE DEPOSITS IN RELATION TO IGNEOUS ACTIVITIES

This chapter deals with the temporal and spatial relationship of secular igneous activity and ore deposits associated with this igneous activity in Japan. It took place during the geotectonic development of the Japanese Islands, and has been classified into the following four groups:

1) Submarine volcanism of basaltic rocks, probably of Late Paleozoic to Early Mesozoic age,

2) Intrusions and extrusions of felsic igneous rocks of Late Mesozoic to Early Tertiary age,

3) Mostly submarine volcanism involving basalt to dacite in Neogene times, and

4) Subareal andesitic volcanism of Quaternary age.

Ore deposits were formed as a result of the igneous activity, and have distinct characteristics with respect to each igneous episode.

9.1 Ore Deposits in Older Rock Terranes

9.1.1 Kieslagers

In the older Japanese rock terranes, particularly in the Sambagawa metamorphic belt, ore deposits are widely developed and have the following characteristics (Fig. 9.1):

1) Massive ore deposits principally consisting of pyrite and chalcopyrite with a little sphalerite and bornite. Ore minerals other than these sulphides are rare.

2) Generally stratiform and harmonious to the country rocks.

3) Embedded in regionally metamorphosed rocks of mafic, pelitic and siliceous compositions. Mafic schists are the most common country rocks.

4) Affected by the metamorphic deformation together with the country rocks.

These ore deposits are often called the Besshi-type deposits, because that found in the Besshi mine, Shikoku, is the most typical. However, they are

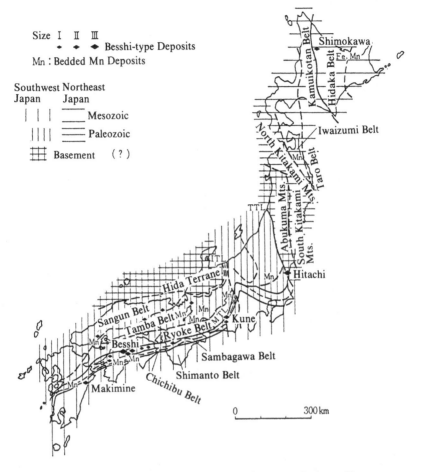

Fig. 9.1. Geotectonic division and the distribution of ore deposits in pre-Neogene terranes (after ISHIHARA, 1978). TTL: Tanakura Tectonic Line, MTL: Median Tectonic Line, ITL: Itoigawa-Shizuoka Line.

sometimes termed "Bedded cupriferous iron sulphide deposits". Incidentally, the geological features described above are similar to those of the Kieslagers developed in the European continent.

In the Sambagawa metamorphic belt this type of sulphide ore is mostly embedded in the thick greenschist layer of the main part of the Minawa Formation (Fig. 9.2; KOJIMA et al., 1956). The grain size of the pyrite in the ores changes systematically in relation to the grain size of the associated schists, increasing as the metamorphic grade increases.

The ore body of the Besshi mine is about 3 m in thickness, extending for

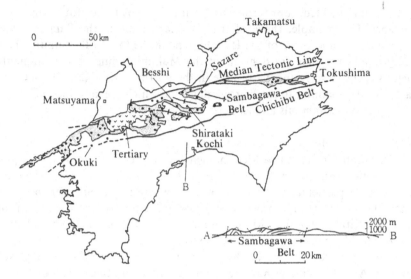

Fig. 9.2. Distribution of Bedded cupriferous iron-sulfide deposits in the Sambagawa Belt in Shikoku (mainly based on Kojima *et al.*, 1956). Gray: the main greenschist rich part of the Minawa Formation, Dot: ore deposit.

about 1,600 m in the direction of strike (E-W), and it is traced for more than 2,000 m in depth. Actually, however, the ore body is separated into two layers with a greenschist bed in between them. The two layers are combined into one layer at its easternmost part. Quartz schist layers envelope the ore body on both sides, and they also are joined together in the eastern part. It is considered, therefore, that the ore body is acutely folded, the crest being to the east, and that its original extent would have been about 3,000 m×2,500 m×1 m, although the body may have subsequently been highly stretched by the metamorphic deformation.

In the Besshi mining district, pyrite is replaced by pyrrhotite at about 1,500 m depth, whereas at depths deeper than 2,000 m, the pyrrhotite-chalcopyrite association is common. Biotite appears in pelitic rocks at the depth which suggests that the metamorphic temperature may have risen.

The geological and petrological features mentioned above indicate the following interpretation for the genesis of the Sambagawa Kieslagers. The present-day characteristics, such as the form of the deposits, the textures and mineral parageneses of the ores are the results of metamorphic deformation and recrystallization. The original rocks would have been formed as sediments deposited in association with the submarine volcanism, through which the original basalts of the associated greenschists would have been made.

Similar bedded cupriferous iron sulfide deposits are found in other terranes, for example, those of the Yanahara mine in the Sangun metamorphic belt, and those of the Hitachi mine in the low P/T Abukuma belt. Additional examples are deposits in the Makimine mine in the Shimanto belt, Kyushu, and in the Taro mine in the northern Kitakami district. The latter is rather similar to the Kuroko deposits, which are described below, in the types of ore minerals and associated igneous rocks.

9.1.2 Bedded manganese ore deposits

A number of small bedded manganese ore deposits occur in the pre-Cretaceous sedimentary terranes (Fig. 9.1). These types of manganese deposits are particularly abundant in areas where numerous chert beds are intercalated with common pelitic and psammitic rocks, for example, in the northern Kitakami, Ashio, Mino-Tamba districts and in the Chichibu belt of the Kanto Mountains, Shikoku and Kyushu.

Of these the deposits of the Ashio district have been especially well studied. About 200 mines have been worked there, but these are mostly very small in size. The following characteristics are observed in the Ashio deposits:

1) The principal ore mineral is rhodochrosite, but tephroite, rhodonite and hausmannite frequently appear in relatively coarse-grained rocks occurring near granite intrusives.

2) Ore bodies are stratiform or lenticular and harmonious to the country rocks.

3) They are generally associated with massive chert, which commonly underlies the ore bodies.

4) The bodies are folded and faulted together with the country rocks.

The original materials for these types of bedded manganese ores are considered to be sedimentary rocks subsequently deposited after the precipitation of massive cherts (WATANABE et al., 1957).

Where manganese deposits occur in high grade metamorphic rocks such as the gneisses of the Taguchi mine district of the Ryoke belt, the main constituents are rhodonite and tephroite.

In the Chichibu and Shimanto belts, bedded ferriferous manganese deposits are known to occur. They consist mainly of Fe and Mn oxide minerals such as manganiferous hematite. Similar Fe-Mn deposits are also found in the Tokoro district, Hokkaido. They are embedded in a particular horizon of stacked pillow basalt layers, and are considered to have been formed in association with submarine volcanism.

9.2 Cretaceous Granitoids and Associated Ore Deposits

The granitoids intruded during Late Mesozoic to Early Tertiary are associated frequently with pegmatites and contact metasomatic or vein-type ore deposits of Sn, W, Mo, Cu, Pb, Zn and Fe (Fig. 9.3). The areal distribution of these deposits is in relation to the types of granitoids associated with them as discussed in Chapter 4 (ISHIHARA and SASAKI, 1973).

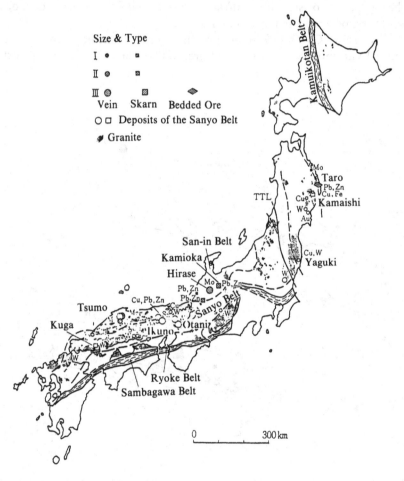

Fig. 9.3. Ore deposits associated with the Cretaceous-Paleogene granitic rocks (after ISHIHARA, 1978). Chromite deposits (Cr) accompanied by ultramafic rocks and Cretaceous bedded sulpphide deposits of the Taro mine (diagonally ruled rhombus) are also shown.

9.2.1 Southwest Japan

The ore deposits of Southwest Japan are generally associated with felsic volcano-plutonic complexes, and rarely with the Ryoke granites which are considered to have been formed at deeper levels.

Molybdenite deposits are found in association with the San-in-Shirakawa zone granitoids, while W deposits are associated with those of the Sanyo-Naegi zone (Fig. 9.4). The Mo deposits are particularly developed in the Daito district, Shimane Prefecture, and commonly consist of molybdenite-quartz veins filling fissures in the granitoids. On the other hand, tungsten deposits, exemplified by those of the Kaneuchi and Otani mines, Kyoto, are scheelite-quartz or scheelite-wolframite-quartz veins in the granitoids or in hornfelses, occurring in the aureoles.

The ratio $Cu/(Zn+Pb)$ in ores also varies systematically with the regional variation of the associated granitoids. The ratio decreases north-

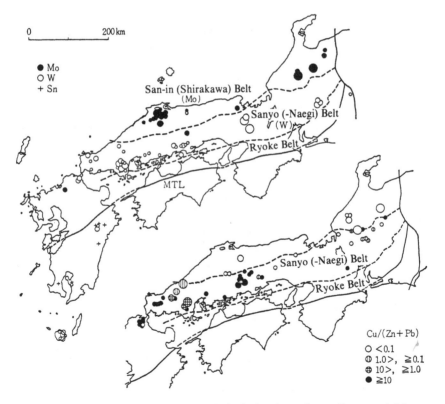

Fig. 9.4. Ore deposits associated with granitoids in Southwest Japan. Upper: molybdenum, tungsten and tin deposits (after ISHIHARA and SASAKI, 1973; TSUSUE, 1976). Lower: Contact metamorphic ore deposits and $Cu/(Zn+Pb)$ of the ores (after SHIMAZAKI, 1975).

wards from 10 to 0.1 in the Sanyo-Naegi zone (SHIMAZAKI, 1975; Fig. 9.4). Contact metasomatic ore deposits rich in Cu are commonly associated with equigranular granites, but those rich in Pb and Zn with granite porphyries and quartz porphyries.

The representative contact metasomatic deposit of Pb and Zn is that of the Kamioka mine, Gifu Prefecture. The mine is located in the Hida gneiss area, and the ore body was formed as a contact metasomatic ore deposit, accompanied by the skarnization of lime-silicate gneisses due to the intrusion of Cretaceous granite porphyries and quartz porphyries.

The Cu, Pb, Zn, Sn and W deposits of the Ikuno and Akenobe mines, Hyogo Prefecture, are considered to have been formed in association with the Cretaceous granite intrusions into Late Paleozoic and Late Mesozoic to Paleogene sedimentary rocks (IMAI et al., 1975). The Ikuno deposit is of the fissure-filling type and includes various minerals such as chalcopyrite, sphalerite, galena, arsenopyrite, pyrite, cassiterire and wolframite. The zonal structure of the deposit, from the inside out, consists of the Sn–W zone, the Cu zone, the Cu–Zn zone, the Zn zone, and further out are the Au–Ag veins (NAKAMURA and MIYAHISA, 1976). The Akenobe deposit, too, is of the vein-type and consists similarly of various minerals. The zonal structure from the inside out comprises the Sn–W zone, the Cu–Sn zone, the Cu–Zn zone and the Zn–Pb zone.

9.2.2   Northeast Japan

Few large ore deposits of Northeast Japan are associated with Cretaceous granites. In the Abukuma plateau, the Cu–W deposit of the Yaguki mine and the Fe–Cu deposit of the Kamaishi mine are mentioned as being representative. The Kamaishi deposit is a contact metasomatic deposit, formed in the skarnized aureole of a Cretaceous granite. The ferriferous ores consist mainly of magnetite, and the cupriferous ores of chalcopyrite and cubanite.

9.3   Neogene Submarine Volcanism and Ore Deposits

A number of ore deposits have been formed in the Neogene Green Tuff Region in association with submarine volcanic activity. Various types of deposit are also developed in Late Cenozoic terranes, other than the Green Tuff Region (Fig. 9.5).

9.3.1   Green Tuff Region
i)   Kuroko deposit

Massive ore deposits, consisting principally of sphalerite, galena and tetrahedrite along with subordinate chalcopyrite, arsenite, scheelite and

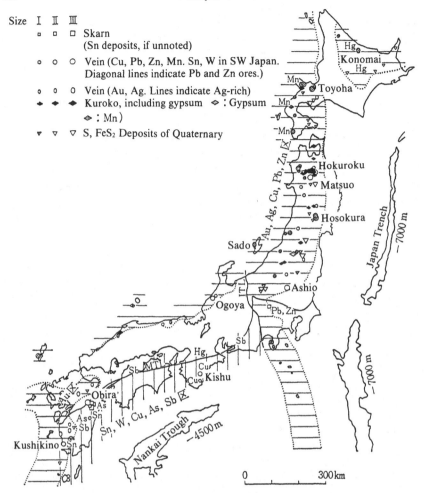

Fig. 9.5.   Ore deposits in Tertiary terranes (after ISHIHARA, 1978). Horizontally ruled: Green Tuff region and Tertiary volcanic areas in southern Kyushu, Vertically ruled: Tertiary volcanic areas in the Outer Zone of Southwest Japan.

quartz, are widely developed in the Green Tuff region of Northeast Japan. These are the Kuroko (black ore) deposits. They have the following characteristics.

1)   Ore minerals

As stated above, they are composed mainly of sphalerite, galena and tetrahedrite, but the other types consisting largely of pyrite and chalcopyrite (yellow ore) and rich in quartz (siliceous ore) are also found. Anhydrite is a fairly common gangue mineral.

2) Geological mode of occurrence

Representative geological structure is illustrated in Fig. 9.6. The yellow and black ore deposits lie successively on a dome-shaped dacitic body, in which the siliceous ore deposits are embodied in places. All of the ore bodies are covered by a thin scheelite layer and further covered by piles of tuff and mudstone. The mudstones are intercalated with thin beds of ferriferous-siliceous sediment. Anhydrite masses are located in distal areas (SATO, 1974). Graded bedding of brecciated ores is observed in some parts of the deposits and called banded ore. (KAJIWARA, 1970).

3) Distribution and age

The Kuroko deposits have been mostly developed in the Neogene Green Tuff Region, and in the Nishikurosawa Stage, of Middle Miocene age (about 15 Ma), when dacitic submarine volcanism was particularly vigorous.

The genesis of the Kuroko deposits has been interpreted from the main features listed above, as follows. Dacitic magma was erupted on to the sea floor and formed a dome-shaped body. At the later stage, the eruption become explosive, and formed breccia around the main body. The hydro-thermal solution containing various metals successively issued forth and precipitated the yellow and black ores. Rocks surrounding the ore body have been silicified into the siliceous ore. After that, all the edifices were covered with mud and tuff. The ore deposits partly collapsed and flowed down the slope of the dome and were deposited as the banded ore. Consequently, the Kuroko deposits are considered to be volcano-sedimentary in origin.

It has been noticed that the Kuroko deposits are developed in areas

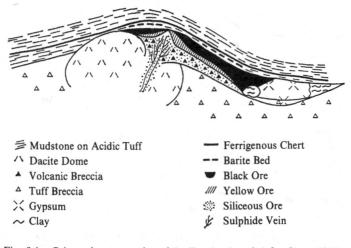

≋	Mudstone on Acidic Tuff	—	Ferrigenous Chert
∧	Dacite Dome	– –	Barite Bed
▲	Volcanic Breccia	▼	Black Ore
△	Tuff Breccia	////	Yellow Ore
✕	Gypsum	⋮⋮	Siliceous Ore
∼	Clay	⅄	Sulphide Vein

Fig. 9.6.   Schematic cross-section of the Kuroko deposit (after SATO, 1974).

having basinal structure such as the Hokuroku and Nishiaizu districts (SATO *et al.*, 1974). In the Hokuroku district in particular, numerous Kuroko deposits are distributed and worked (Fig. 9.7). The total size of the Kuroko ore mass, whose copper content is about 2%, is estimated to be about 1,000 million tons up to the present.

ii)   *Vein-type and contact metasomatic ore deposits*

A number of vein-type ore deposits of various metals such as Cu, Pb, Zn, Au and Ag occur in the strata younger than those of the Kuroko

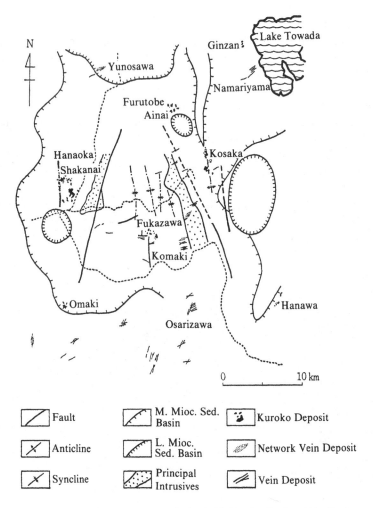

Fig. 9.7.   Geologic structure and ore deposits in the Hokuroku district, Akita Prefecture (after SATO *et al.*, 1974).

deposits. The vein-type deposits would have mostly formed in the Nishikurosawa and Funakawa Stages (15 to 7 Ma). The Toyoha deposit (Pb, Zn) in Hokkaido, the Hosokura (Pb, Zn), Osarizawa (Cu) and Ani (Cu) deposits in the Tohoku district and the Ashio deposit (Cu) in the Kanto area are the representative. The Au–Ag deposit of Konomai, Hokkaido, and those of the Sado and Izu districts, and the Hg deposit of Itomuka, Hokkaido, should also be mentioned.

The Asio copper mine was previously the largest in Japan, and yielded about 800,000 tons of Cu ore until it was shut down in 1973. The deposit is of the vein-type and occurs in veins in the Asio rhyolitic welded tuff and in the surrounding Jurassic sediments. They are similar to the deposits of the Ikuno-Akenobe area mentioned above. The zonal structure from the inside out consists of the Sn–W–Bi–Cu, the Cu–As–Zn and the Zn–Pb– Cu–As zones.

In the Chichibu district of the Kanto Mountains, where the Paleo-Mesozoic sediments are widely exposed, the contact metasomatic ore deposit of the Chichibu mine occurs. It is developed in association with skarnized limestone, and mainly consists of magnetite, and is also accompanied by vein-filling auriferous Pb–Zn deposits.

9.3.2 Neogene metallogenic districts other than Green Tuff Region

Numerous Au–Ag deposits are developed in northern and southern Kyushu, where Neogene volcanic rocks are widely exposed. They are the Au–Ag-bearing quartz veins filling fissures in altered andesites and rhyolites, exemplified by deposits of the Taio mine, northern Kyushu, and of the Kushikino and Kasuga mines, southern Kyushu. The last deposit contains native sulphur, pyrite, enargite and covelline in addition to native gold, and is successively surrounded outwards by the silicified zone, the alunite zone, the dickite zone and finally by the limonite-kaoline zone.

Along the Median Tectonic Line in Southwest Japan, a few Sb or Hg deposits are developed in association with Neogene igneous rocks: for example, the Ichinokawa Sb mine in Ehime Prefecture, the Tsugu Sb mine in Aichi Prefecture and the Yamato Hg deposit in Nara Prefecture.

Furthermore, the Neogene granitic masses in the Outer Zone of Southwest Japan are accompanied with a few vein-type or contact metasomatic ore deposits. The representatives are the Kishu deposit of central Kii Peninsula and the Obira deposit in central Kyushu. Both are polymetallic mainly of Cu, Pb, Zn, Sn and As. The Obira deposit is well known for yielding good crystals of axinite and danburite.

9.4 Sulphur and Pyrite Deposits in the Quaternary Volcanic Districts

Native sulphur and pyrite have been occasionally precipitated in association with Quaternary subareal volcanic rocks. More than 100 deposits were being worked by 1955, and the annual output of sulphur and pyrite were 200,000 tons and 1 million tons respectively. However most of the workings have now been shut down.

The principal mining districts are, from north to south, the Shiretoko-Akan, Toya, Oshima Peninsula, Shimokita Peninsula, Hachimantai, Narugo, Zawo-Azuma, Joshin-etsu and Kuju (Fig. 9.5). The deposits were formed through sublimation from volcanic gas, precipitation from similar gas issuing from ponds and lakes and replacement of rocks by hydrothermal solution.

The Matsuo deposit in the Hachimantai area was the largest in Japan. It is developed in strata composed of andesitic lavas, tuff breccias and tuffs and extends about 1.45 million square meters in area, and about 80 m in thickness. The principal ore minerals are native sulphur and iron sulphides. The country rocks of the deposit are strongly altered and show a zonal structure consisting, from the inside to outside, of sulphur deposits, pyrite deposits, the opal zone, the alunite zone, the kaoline zone and the saponite zone.

The sulphur deposits occur commonly at shallow depths. It is inferred that groundwater would have been circulating there when the deposits were precipitated from the pneumatolitic and hydrothermal fluids (MUKAIYAMA, 1959).

Finally, a peculiar sulphur deposit should be mentioned. Melted sulphur flowed down the northern slope of the Io-san Volcano in the Shiretoko Peninsula, Hokkaido, during 1876–1889 and in 1936. In the latter case, about 150,000 tons of sulphur issued forth for eight months. The deposit has already been mined out (WATANABE, 1940).

REFERENCES

IMAI, H., LEE, M. S., IIDA, K., FUJIKI, Y., and TAKANOUCHI, S., Geologic structure and mineralization of the xenothermal vein-type deposits in Japan, *Econ. Geol.*, **70**, 647-676, 1975.

ISHIHARA, S., Metallogenesis in the Japanese island arc system, *J. Geol. Soc. Japan*, **135**, 389-406, 1978.

ISHIHARA, S. and SASAKI, A., Metallogenic map of Japan, Plutonism and mineralization, (1) Molybdenum, Tungsten and Tin, 1/2,000,000, Geol. Surv. Japan, 1973.

KAJIWARA, Y., Syngenetic features of the kuroko ore from the Shakanai mine, *Volcanism and Ore Genesis* (T. Tatsumi ed.), Univ. Tokyo Press, 197-206, 1970.

KOJIMA, G., HIDA, K., and YOSHINO, G., The stratigraphical position of Kieslager in the Sambagawa crystalline schist zone in Shikoku, *J. Geol. Soc. Japan*, **62**, 30–45, 1956.

MUKAIYAMA, H., Genesis of sulphur deposits in japan, *J. Fac. Sci., Univ. Tokyo, I*, **11**, Suppl., 1–148, 1959.

NAKAMURA, T. and M. MIYAHISA, Hypogene zoning and role of geologic structures of vein-type deposits, *mining Geol., Spec. Issue*, **7**, 33–57, 1976.

SATO, T., Distribution and geological setting of the kuroko deposits, *Mining Geol. (Japan), Spec. Issue*, No. 6, 1–9, 1974.

SATO, T., TANIMURA, S., and OHTAGAKI, T., Geology and ore deposits of the Hokuroku district, Akita Prefecture, *Mining Geol. (Japan), Spec. Issue*, No. 6, 11–18, 1974.

SHIMAZAKI, H., The ratio Cu/Zn+Pb of pyrometasomatic deposits in Japan and their genetical implications, *Econ. Geol.*, **70**, 717–724, 1975.

WATANABE, T., Eruption of molten sulphur from the Shiretoko Iwo-san volcano, Hokkaido, Japan, *J. Geol. Geogr.*, **17**, 289–310, 1940.

WATANABE, T. *et al.*, Geological map of the Ashio district and the explanation text, Tochigi Prefecture Office, 1957.

CHAPTER 10

GEOTECTONIC HISTORY OF THE JAPANESE ISLANDS

10.1 Middle Paleozoic

The oldest strata of the Japanese Islands are of Silurian age. In Southwest Japan Silurian rocks are exposed in the Hida Border Fault Zone and in the Kurosegawa Zone, and are accompanied by Devonian rocks. The Siluro-Devonian rocks are also distributed in the Nagasaka district of the southern Kitakami Mountains and in the Matsugadaira area of the Abukuma Plateau, both in Northeast Japan.

Lavas and pyroclastic rocks, often welded tuffs, of rhyodacitic composition of the calc-alkali rock series are characteristic of the Japanese Siluro-Devonian formations, in addition to the limestones, mudstones and sandstones of warm and shallow sea environments. The lithofacies suggests that the strata would have been formed in regions such as island arc and continental margin regions, which were underlain by sialic crust of more than 18 km in thickness.

Much debate has taken place on the geotectonic history of the Siluro-Devonian terranes (ICHIKAWA et al., 1972; KIMURA, 1977; HADA et al., 1979; SUZUKI et al., 1979). As stated in Chapter 2, most of these terranes consist not only of the Siluro-Devonian strata, but of various types of rocks such as high-temperature metamorphic rocks, high-pressure metamorphic rocks, sheared granites and serpentinites, and are considered to be serpentinite melange zones (MARUYAMA et al., 1984). It may be that the Siluro-Devonian rocks were developed as accretional terranes at the boundary between the Asian Continental and Pacific Oceanic plates (Fig. 10.1). (HORIKOSHI, 1972; SAITO and HASHIMOTO, 1982; MARUYAMA et al., 1984).

10.2 Late Paleozoic

Strata consisting of slate, sandstone, chert, limestone and greenstone, which are the most widely exposed of the Japanese Pre-Cretaceous rocks,

Fig. 10.1. A model showing the geotectonic development of the Japanese Siluro-Devonian terranes. av: Acid to intermediate volcanic rocks of the calc-alkali series, ls: limestone, O and R: Omi and Renge metamorphic rocks, Mo and Ma: Motai and Matsugadaira metamorphic rocks.

were for long time considered to be of Permo-Carboniferous age. However recent micro-fossils studies indicate that the strata are substantially of Triassic to Jurassic age, and Late Paleozoic rocks are apparently rather scarcely distributed.

Permo-Carboniferous strata are exposed in two zones. One extends from the Chugoku, through Tamba-Mino to Ashio districts on the inner (Sea of Japan) side, while the other is in the Chichibu belt on the outer (Pacific) side.

Excluding those in the southern Kitakami Mountains and the Abukuma Plateau, the principal rock types are greenstone, limestone, chert and slate. Limestone in particular is predominant in the Permo-Carboniferous strata of Southwest Japan, in close association with greenstone, the former resting on the latter. Greenstone derived from basalt is often highly amygdaloidal and suggests that the eruptions took place in a fairly shallow sea environment. Limestone is also of shallow-sea origin, and contains abundant fossil reef corals, algae and oolites. Limestone plateaus such as those of Akiyoshi, Taishaku, Atetsu and Omi are reef complexes which developed on basaltic volcanic bodies, the basaltic sub-stratum having been truncated at less than 400 m depth. A part of the original basalt forming the greenstones was often alkalic, and it is suggested that the greenstone bodies represent the upper parts of seamounts formed on the flanks of mid-oceanic ridge.

Permian limestone masses are also developed in such mountainous areas as Ibuki, Fujiwara, Funabuse and Ashio in the Inner Zone and as Tsukumi, Onohara and Bukozan in the Outer Zone. They are also shallow-sea reef complexes resting on greenstone bodies. In the Ashio district the complex is embedded in the Jurassic formation as an allochthonous mass.

Permian strata consisting of greenstone, chert and slate are exposed in the Chugoku, Tamba-Mino and Ashio areas of the Inner Zone and in the Chichibu belt and the northern Kitakami Mountains of the Outer Zone. Some of the greenstone masses are considered to be ophiolitic complexes, whose representative is the Yakuno ophiolite in the Maizuru Zone

Fig. 10.2. Geotectonic development of Southwest Japan in Late Permian. av: Acid to intermediate volcanic rocks of the calc-alkali series, Ak: Limestone masses of Akiyoshi, Atetsu, Omi, etc. The Chichibu terrane is considered to have been a micro-continent derived from Pacifica.

(ISHIWATARI, 1985).

In the Sangun high P/T metamorphic belt there are many mafic to ultramafic bodies, which comprise various types of rocks such as diabase, gabbro, serpentinite and trondhjemite. Although the bodies are thrusted over the Sangun terrane, they seem to have been derived from oceanic crust underlying the sediments, to become the Sangun metamorphic rocks. Gabbroic rocks, dated as 406–424 Ma in the Nagato Tectonic Zone, and as 324–379 Ma in northern Kyushu are probably of oceanic crustal origin. Similarly, the original sedimentary strata of the Sambagawa high P/T belt seem to have been underlain by oceanic crust (IWASAKI, 1979) (Fig. 10.2).

10.3 Early Mesozoic

The following five zones, from north (inner) to south (outer), are classified in the Triassic strata of Southwest Japan.

(1) Morasse-type sediments of Middle to Upper Triassic age, in the Atsu-Miné formations of western Chugoku. A number of coal seams are intercalated. The lithofacies suggests deltaic or floodplain environments.

(2) Lower and Upper Triassic strata of the Maizuru Zone. The lower Triassic strata change from near-shore coarse-grained sediments in the northwestern area to offshore, fine-grained sediments in the southeastern area, indicating a submarine fan environment. On the other hand, the Upper Triassic strata are lithologically similar to sediments of the Miné Formation mentioned in (1).

Both (1) and (2) are never associated with chert and greenstone.

(3) Chert-dominant Triassic strata are widely distributed in the southern Chugoku, the Tamba-Mino-Ashio terranes, the marginal part of the Ryoke belt, the southern part of the Sambagawa Belt and the northern part of the Chichibu Belt.

(4) Shallow-sea limestones and thick cyclic piles of clastic sediments occur in a few areas near the Kurosegawa Zone. They are considered to be

continental slope deposits.

(5)   Strata of chert and slate and of greenstone, chert and micritic limestone. The former are distributed in the northern part of the Sambosan belt and the latter in the southern part of the Sambosan belt to the south of the Chichibu belt. Furthermore, the latter strata comprise vesicular pillowy basalt, hyaloclastite and tuff and, in part, talus deposits carrying blocks of reef limestone, greenstone and chert. Strata similar to the former are also exposed in the northern Kitakami Mountains and those to the latter in the Iwaizumi district, both in northeast Japan.

Basaltic rocks in the Sambosan-Iwaizumi zone are considered not to be of island arc but oceanic type basalt. Therefore, the strata of this zone would have been deposited in shallow submarine volcanic areas far from continents. The imbricate structure of the Sambosan belt, due to a number of northward dipping thrusts, is interpreted to be an accretionary prism formed in Middle to Late Jurassic time (HADA et al., 1979).

Additional Triassic events which should be mentioned are: (a) The intrusion of the Funatsu granites and associated low P/T metamorphism in the Hida terrane, and (b) the high P/T Sangun metamorphism.

Radiometric ages of the Hida metamorphic rocks are mostly grouped into three age clusters of 230–250 Ma, 210–220 Ma and 170–180 Ma. The last corresponds to that of the Funatsu granites. The Funatsu granites were intruded into the Upper Carboniferous rocks, and the Hida gneisses are overlain by the Middle Jurassic Tetori Group. These data indicate that the principal phase of the Hida metamorphism is of Late Permian to Triassic age and the intrusion of the Funatsu granite of Triassic to early Jurassic age.

The Sangun rocks are covered by the Upper Triassic sediments, so that metamorphism would have taken place earlier than Early Triassic. The Sangun Belt is considered to constitute a pair with the Hida Belt terrane (MIYASHIRO, 1967).

Distinct asymetrical variation is observed in sedimentary and meta-morphic facies of the Triassic rocks, particularly those in Southwest Japan. Figure 10.3 gives a general view of the geologic history during the Mesozoic of Southwest Japan.

10.4   Middle Mesozoic

Jurassic strata are classified into the following seven zones from the Sea of Japan to Pacific sides.

(1)   Non-marine to shallow-sea Morasse-type sediments of the Hida and its border areas.

(2)   Shallow-sea thick cyclic sediments of the Sangun-Chugoku belts.

(3)   Marine flysch-type sediments, rich in pale-green tuff and sandstone

of the Tamba-Mino-Ashio terranes.

(4)   Shallow-sea thick cyclic sediments in the Kurosegawa Zone, the eastern marginal zone of the Abukuma plateau and the southern Kitakami Mountains.

(5)   Marine flysch-type sediments, rich in pale-green tuff and sandstone of the northern Sambosan belt.

(6)   Shallow-sea limestone and clastic sediments of the southern Sambosan belt.

(7)   Marine Flysch-type sediments of the northern Shimanto belt.

In contrast to the Triassic rocks, sandstone is predominant in the Jurassic strata, and olistostromal deposits, carrying blocks of Paleozoic greenstone, limestone and Jurassic chert, are often developed.

The Lower Jurassic Kuruma Group unconformably overlies the metamorphic rocks and serpentinite of the Hida border belt. On the other hand, the Tetori Group of Middle Jurassic to Lower Cretaceous covers the Hida gneisses and associated Funatsu granites. The latter group frequently contains pebbles of orthoquartzite, probably derived from the Asian Continent. Both groups show considerable local variation in lithofacies and thickness, suggesting violent crustal fluctuations during that period.

Thick mudstone-sandstone turbidites with pale-green tuffaceous shale cover the Lower to Upper Triassic strata in the Mino area. They seem to be of Middle Jurassic to Earliest Cretaceous in age, based on radiolarian remains. The turbidites are, in some places, interlayered with olistostromal beds bearing large blocks of Permian fusulinid limestone, greenstone and Triassic chert. It is suggested that the rapid uplifting of the provenance area occurred, this is probably located to the north of the Mino area (ADACHI, 1976, 1979).

The stacking of a number of thrusted sheets, amounting to a few hundred meters in thickness, is observed in the Triassic chert and Jurassic tuff and turbidite formations. The sheets are piled up parallel to each other (IGO, 1979). The structure would have been constructed by low-angle, or bedding, thrusting which occurred along the basal planes of the Triassic strata. The thrusting movements resulted in considerable shortening, in areal extent, of the strata. The thrusted sheets are further affected by upright folding and covered by Late Cretaceous Nohi Rhyolites, suggesting that the folding took place in Early Cretaceous time (KIMURA, 1960) (Figs. 10.3(b) and (c)).

Addtional events of the Jurassic time were the Ryoke-Abukuma low P/T regional metamorphism and associated plutonism, and the Sambagawa high P/T regional metamorphism should also be mentioned. On the other hand, ophiolitic formations were developed in the Central and Tokoro belts of Hokkaido.

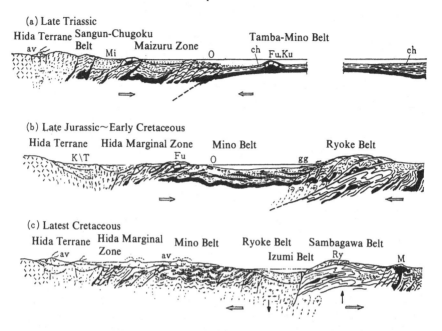

Fig. 10.3. Mesozoic geotectonics of Southwest Japan. (a): Mi: Atsu and Mine groups, Tl; Triassic limestone masses of Taho, Kamimura, etc., Ko: Triassic clastics of the Kochigatani Formation, etc., Fu, Ku, etc.: Early to Middle Permian limestone of Funabuse-yama, Kuzuu, etc., ch: Chert, rm: Micritic limestone rich in radiolaria, av: Acid to intermediate volcanic rocks

The Ryoke and Sambagawa metamorphic belts form a pair; however, its elongation trend is oblique to that of the Sangun-Chugoku, Maizuru and Tamba-Mino-Ashio terranes to the north. As a result, the Sangun terrane is in direct contact with the Ryoke terrane in Kyushu, without the Jurassic turbidite-flysch facies in between. However, in central Japan, the Jurassic strata are widely distributed to the north of the Ryoke belt. The configuration of these terranes suggest that subduction of the Sambagawa belt was oblique to the Jurassic terranes of the Inner Zone and that a basin formed on the landward side.

The original materials of the Sambagawa metamorphic rocks consist of mafic to ultramafic rocks, and clastic sediments derived from them. Also trench-filling ordinary clastic sediments were added to them.

The Middle to Upper Jurassic strata, distributed in areas near the Kurosegawa Zone, are shallow-sea sediments of the thick cyclic facies type deposited on the shelf, and contain oolite and reef limestone rich in colonial corals and algae. Similar Jurassic rocks are also exposed in the eastern

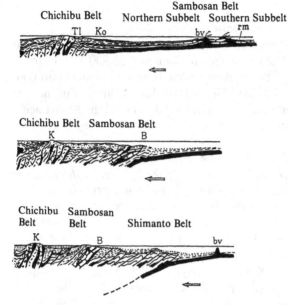

of the calc-alkali series, O: Olistostrome. (b) K/T: Kuma/ Tetori Group, B: Butsuzo line, gg: Gravity sliding, K: Kurosegawa Zone. (c) Ry: Remnant part of the thrusted Ryoke terrane, M: Mikabu ophiolite.

Abukuma Plateau and southern Kitakami Mountains. Owing to subduction of the Sambagawa belt, the Chichibu belt at that time, would have come close to the Ryoke belt, from which clasts of chert in the Jurassic strata were derived.

10.5 Late Cretaceous in Southwest Japan

The Cretaceous strata of Southwest Japan are classified as follows.

(1) Non-marine, thick cyclic sediments of lowermost Cretaceous age, and intermediate to acidic volcanic rocks of Middle to Upper Cretaceous age in the Hida, Sangun-Chugoku, Tamba-Mino-Ashio terranes.

(2) Upper Cretaceous Morasse-type sediments are distributed in the Izumi terrane between the Ryoke and Sambagawa belts.

(3) Lower to Upper Cretaceous thick cyclic sediments are exposed in the vicinity of the Kurosegawa Zone.

(4) Lower to Upper Cretaceous flysch-type sediments and underlying

pre-flysch deposits occur in the northern part of the Shimanto belt.

10.5.1 Intermediate to acidic volcanism of Inner Zone

Four cycles of acidic volcanic activity were analysed: the earlier two being of Cretaceous age. The ejecta amount to about 51,900 km^2 in total, including those of the Abu, Takada, Aioi-Arima, Ikuno, Nohi and Okunikko districts (ICHIKAWA *et al.*, 1968). This tremendous acid volcanism was considered to have occurred in relation to the subduction of the Kula-Pacific Ridge (UYEDA and MIYASHIRO, 1973).

10.5.2 Retroduction and uplift of the Sambagawa belt, fracture and subsidence of the frontal part of the Ryoke belt, and sedimentation of Cretaceous strata in the border area between the two belts.

Upper Cretaceous strata are distributed in a narrow zone between the Ryoke and Sambagawa Belts. Considerably thick (15 km in the maximum thickness) piles of sediments rich in coarse clastics were deposited for a very short period. The main sedimentation area moved eastwards from the Goshora Group in Western Kyushu through to the Onogawa Group in Shikoku, and to the Izumi Group (post Campanian). It is due to the eastward migration of the retroduction of the Sambagawa belt, which resulted in fracturing and subsidence of the frontal part of the Ryoke belt (TAIRA *et al.*, 1979). This is in agreement with the age of granite intrusions in the Ryoke belt which also become younger eastwards (MATSUMOTO, 1977).

10.5.3 Formation of Median Tectonic Line

The Median Tectonic Line may be interpreted as an up-thrusting fault of the Ryoke belt onto the Sambagawa belt, having been caused by the retroduction of the latter.

10.6 Development of Shimanto Belt

10.6.1 Orogenic polarity of the Shimanto Belt

In the Shimanto belt, strata become younger from north to south: Uppermost Jurassic to Cretaceous in the northern sub-belt, Oligocene to Lower Miocene in the southern part of the southern sub-belt (HADA *et al.*, 1979). However, because the strata generally dip northwards, they become younger to the north in each sector. Furthermore, the deposits unconformably overlying the folded and thrusted strata also become younger from north to south and further to the submarine regions (Fig. 10.4).

10.6.2 Tectogenetic process of the Shimanto belt

The strata of the Shimanto belt are classified, independent of their ages,

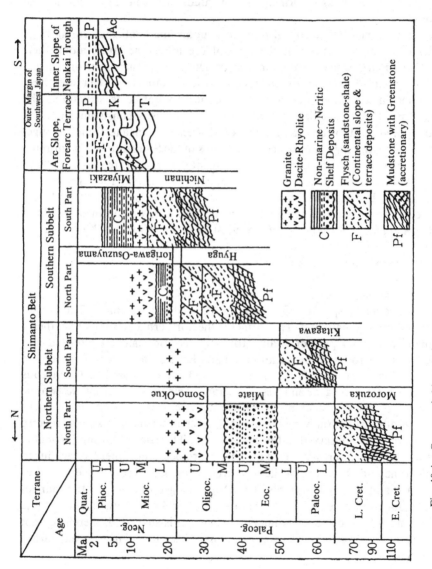

Fig. 10.4. Geotectonic history of the Shimanto belt and the submarine terranes further to the south (after OKUDA *et al.*, 1979). T, K, Ac and P: submarine strata.

into the following two categories:

(1)   Strata consist of intercalations of sandstone and shale (Flysch facies).

(2)   Strata consist principally of mudstone containing greenstone blocks.

All strata appear to lie conformably one upon one another, forming huge piles of strata. However, at the base of the mudstone layer, along their southern margin, northward dipping thrusts occur. Therefore, one thrust sheet consists of an upper flysch layer and a lower mudstone layer, and these combine in a number of sheets to produce an imbricate structure. The mudstone layers include some olistostromal deposits consisting of many blocks of greenstone and sandstone, which seem to form a tectonic melange.

The stratigraphic and structural features of the Shimanto belt mentioned above are characteristic of an active continental margin or trench-arc system. The mudstone layers may be accretionary prisms, and their greenstone blocks may be fragments of seamounts. On the other hand, the flysch layers are comparable with sediments in fore-arc basins or on continental slopes. (HADA *et al.*, 1979; TAIRA *et al.*, 1979; SUZUKI and HADA, 1979).

10.7   Pre-Neogene of Hokkaido

10.7.1   Orogenic polarity

Geologically speaking, Hokkaido is divided into two parts: The Southwestern and the main part. The former is the northern extension of northeast Japan. The axial part of Main Hokkaido is underlain by plutonic and metamorphic rocks of the Kamuikotan and Hidaka Belts. Strata to the west of the axial zone, mostly dip eastwards and are composed of Cretaceous (Yezo Group), Paleogene and Neogene strata occurring successively westwards. Therefore, reverse stacking is observed here, that is, the older rocks structurally overlie the younger rocks. On the other hand, a similar reverse structure is also observed in the eastern area. The Hidaka Group (probably Jurassic), Upper Jurassic strata of the Tokoro belt and Upper Cretaceous to Paleocene and Oligocene of the Nemuro Group are successively exposed to the east, and all dip westwards. Therefore, the geology of the Main part of Hokkaido is symmetrical in appearance, but rocks on both sides are quite different in lithology and sedimentary facies. It is suggested, therefore, that two geologically different masses would have been united in the axial part of the island.

10.7.2   Kamuikotan and Hidaka metamorphic belts

In the axial part of Hokkaido, the Kamuikotan high P/T and the Hidaka low P/T metamorphic belts run side by side in a N-S direction. But, in contrast to most of the paired metamorphic belts in the circum-Pacific regions, the high P/T belt is located on the continental side and the low P/T belt on the oceanic side.

The Kamuikotan high P/T rocks are overlain by the ophiolite complex of the Sorachi Group, and mostly consist of oceanic materials, without terrestrial sediments. Radiometric ages measured to date are 109 Ma and 120 Ma. On the other hand, the Hidaka low P/T rocks are considerably younger, and the plutonic rocks have been dated as 20 Ma to 30 Ma. Therefore, the two metamorphic belts are not considered to constitute a pair.

According to recent detailed studies, the western part of the Hidaka belt is an overturned ophiolitic complex. However the main part represents a cross-section of continental or island-arc crust, consisting, from lower to upper, of peridotites, granulites, amphibolites, gabbro, gneisses, hornfelses and finally unmetamorphosed sedimentary rocks (KOMATSU et al., 1979).

10.7.3   History

The Cretaceous Yezo Group, exposed in areas to the west of the axial zone, rapidly thins out towards the Ishikari Lowland. Materials of the Yezo Group would have been supplied from the west by a volcanic zone of acid-intermediate rocks of Late Jurassic to Early Cretaceous age. The accompanying trench may have been located to the east of the Yezo Group sedimentary basin (Fig. 10.11) (OKADA, 1979; DICKINSON, 1978).

The Kamuikotan belt, where high-pressure metamorphic rocks are overlain by low-pressure metamorphosed ophiolite, would have formed a structural high at the trench-slope break (OKADA, 1979). Accordingly, the Yezo Group to the west and the east of the Kamuikotan belt are quite different in sedimentary facies: in the west thick coarse clastics occur, while in the east mostly fine mudstones occur.

On the other hand, clastic materials of the Yubetsu, Nemuro and Hidaka Groups exposed in areas to the east of the axial zone would have come from the northeast (Okhotsk Paleo-continent) (KIMINAMI et al., 1978). A part of the Hidaka Group to the west of the Hidaka metamorphic belt is rich in mudstone, and also includes chert and radiolarian micritic limestone. Furthermore, greenstone and serpentinite are frequently associated with them. It is suggested that at that time, sediments in the frontal areas of the Okhotsk Continent would have spread over the oceanic crust. Recent detailed studies have shown that the principal part of the Hidaka metamorphic belt formed part of the continental or island-arc crust (KOMATSU et al., 1979; DICKINSON, 1978).

The axial zone is underlain by ophiolites of the Kamuikotan belt, and ophiolites of the westernmost and western sub-belts of the Hidaka belt, in association with chert and micritic limestone. The rocks represent an oceanic environment (the so-called Paleo-Kamuikotan Sea) at that time (HORIKOSHI, 1972). By the beginning of Paleogene, this oceanic area would have been consumed, due to westward subduction, caused by westward migration of the Okhotsk Continent rotating clockwise, and eventually the two continental blocks collided at the present-day axial zone. Consequently, the sedimentary basin was split and migrated both ways to the western Ishikari and the eastern Nemuro terranes. Subsequent uplifting of the axial zone caused further east and west migration of the sedimentary basins, where thick Neogene strata, including Morasse-type sediments, were deposited.

10.8   Northeast Japan in the Late Cenozoic

The island arc systems fringing the Northwest Pacific, generally consist of an outer non-volcanic and an inner volcanic arc. In the Northeast Japan Arc, the area extending from Kitakami to Abukuma forms its outer arc, while the Backbone Range and the contiguous regions to the west represent the inner arc (Fig. 10.5). Similarly, the region extending from the Habomai Islands through the Nemuro Peninsula to Kushiro in Hokkaido, is the outer arc of the Kurile Arc, and the volcanic zone of Etorou-Kunashiri-Shiretoko-Tokachi is the inner arc. The volcanic front located at the outer margin of the inner arc has hardly changed its position since Early Miocene times. The geologic structures formed since the Miocene in Northeast Japan are generally parallel to the volcanic front, but are oblique to the structures of the pre-Tertiary basement rocks. It is suggested that a continual series of crustal movement has occurred since Early Miocene time. (MATSUDA et al.,

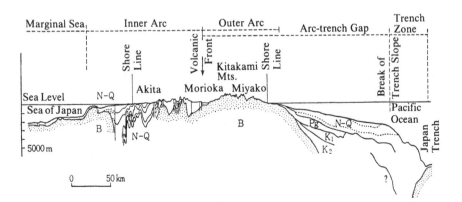

Fig. 10.5.   E-W section of the Northeast Japan Arc (modified after ISHIWADA et al., 1977). Aspect ratio is 10:1. N-Q: Neogene-Quaternary, Pg: Paleogene, K_1: Upper Cretaceous, K_2: Lower Cretaceous, B: Basement rocks.

1967; SUGIMURA *et al.*, 1963).

In central Hokkaido and southern Fossa Magna, the late Cenozoic structural trends cross the volcanic front at high-angles. In the westernmost part of the Kurile Arc and the northernmost part of the Izu-Ogasawara Arc, volcanoes are located in a few en echelon zones (Fig. 10.6).

The area outside the front (the outer arc) region uplifted in the Cenozoic times. The Kitakami Mountains and Abukuma Plateau were upheaved, and the Abukuma Plateau is a tilted block. The Kitakami Mountains uplift produced low-relief erosion surfaces, and raised marine terraces are also observed. The comparatively small scale crustal movements, such as folding of strata, did not take place there, excepting in areas near large basement faults (Futaba Fault and Tanakura Shear Zone). In the inner volcanic belt of Northeast Japan, on the other hand, the vertical crustal movement were highly variable in time and space, and consequently, the thickness of sediments and the resultant topography is also variable. The present day altitude of the basal unconformity developed in the Miocene, shows the sum of vertical movement since Miocene times. As shown in Fig. 10.7, in the

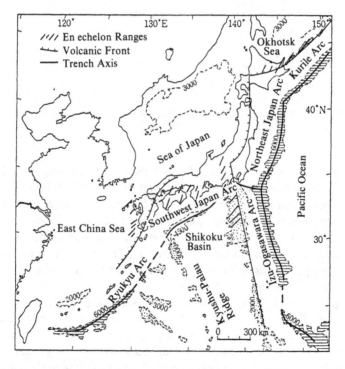

Fig. 10.6.   Arcs, trenches, volcanic fronts and en echelon ranges both on and offshore the Japanese Islands (after KAIZUKA, 1975 and others).

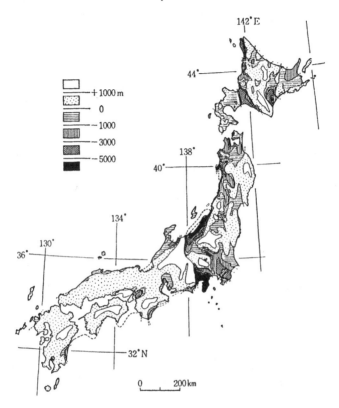

Fig. 10.7.   Uplifing and subsidence since Miocene times (after MATSUDA *et al.*, 1967).

Inner Zone of Northeast Japan, 1) subsidence is remarkable, 2) its degree is highly variable from place to place, and 3) it has a relatively large scale in the Sea of Japan coast area (more than −6,000 m in the Niigata oil field).

The main subsidence took place in the Miocene, while the main uplift was in Plio-Pleistocene times. The earliest marked subsidence in Miocene times occurred in the axial mountain area, but this soon changed into uplift. The subsiding area has migrated westwards to the Sea of Japan (KITAMURA, 1959), where subsidence is still going on.

The Late Cenozoic strata filling the subsidence areas have been more or less folded (Fig. 10.8). The folded structure was mostly developed after deposition of the Pliocene. In the Sea of Japan coastal area the folding movement is active even today causing deformation of river terraces and shifting of benchmarks. The rate of inclination of the fold limbs, which are a few kilometers in wave-length, is estimated as $n \times 10^{-7}/y$ (KAIZUKA, 1968).

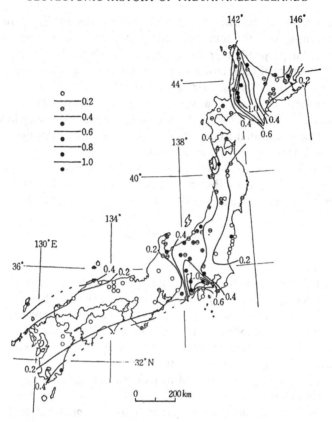

Fig. 10.8. Grade of crustal deformation since Miocene times (after MATSUDA et al., 1967). Numerals indicate the ratio of vertical displacement/width of the respective cross-section.

On the other hand, the Quaternary active faults are mostly reverse faults, and are parallel to the volcanic front, their average slip rate is about $n \times 10^{-1}$ mm/y (RESEARCH GROUP OF ACTIVE FAULTS, 1980). These features suggest that during latest Neogene to Quaternary times, Northeast Japan experienced compressional tectonic conditions, whose maximum compression axis lies in a WNW direction, roughly perpendicular to the island arc.

According to NIITSUMA (1979), the rate of sedimentation was considerably less in the Late Miocene (4.7–10.4 Ma), and the relief of mountains was very low, and wide erosional surfaces were formed. However in Middle Miocene times, (10–16 Ma) the rate was moderate, the sea area very wide, and bimodal submarine volcanism lead to the formation of the Kuroko deposits. Accompanying the volcanism, crustal subsidence and faulting commenced in areas which had been land during the Middle Miocene, the sedimentary basin becoming gradually deeper. It is suggested that Northeast

Japan experienced extensional tectonic conditions in Early to Middle
Miocene times, with the minimum compressional axis having been in an
E-W direction perpendicular to the arc trend (NAKAMURA and UYEDA,
1980). The history of the tectonic conditions prevailing in northeast Japan in
Late Cenozoic times is summarized as follows.

(1)   Extensional, but relatively low in rate of deformation in Miocene.
In the later period (10–4.7 Ma), the sedimentation rate was very low.

(2)   Compressional tectonics with high rate of deformation occurring
at the end of Tertiary to Quaternary time.

The geological history may be broadly interpreted as the growth and
breakdown of the plate subducting into the deeper mantle (KANAMORI,
1977) since Middle Miocene times (NIITSUMA, 1979) (Fig. 10.9).

10.9   Southwest Japan in the Late Cenozoic

Two distinct structural trends are observed in the Late Cenozoic of
Southwest Japan. One trend is parallel to the elongation of the Nankai
Trough, while the other trend crosses it obliquely (Fig. 10.10). The former is
related to the crustal movements, whose wavelengths are comparatively long.
It is comprised of the following zones, from north to south:

(1)   Subsidence zone of the coastal areas of San-in to Hokuriku, where
volcanic activity has occurred since the Miocene,

(2)   An uplifting zone, extending from the Chugoku through Tamba to
the Mino-Hida districts,

(3)   Subsidence zone of the Setouchi area to the north of the Median
Tectonic Line,

(4)   An uplifting zone in central Kyushu-Shikoku-Kii, and

(5)   Subsidence zone extending further south to the Nankai Trough.
The present-day relief of these zones, however, has been mostly formed
during the Quaternary. For example, about 1,000 m of the present day
altitude of 1,200–1,900 m of the mountainous outer zone (4) was attained in
the Quaternary. This type of crustal movement appeared embryonally since
Early or Middle Miocene times. The total vertical displacement since the
Miocene in Southwest Japan is less than that of the same period in Northeast
Japan (Fig. 10.7).

One of the distinctive features, parallel to the zonal structure of
Southwest Japan, is the Median Tectonic Line, which caused the right-
lateral displacement in Late Quaternary times at the rate of 0.5–1 cm/y
(OKADA and ANDO, 1979). This is considered to have resulted from the
oblique subduction of the Philippine Sea plate under Southwest Japan
(FITCH, 1972).

The other structural trend characteristic of the Quaternary in Southwest

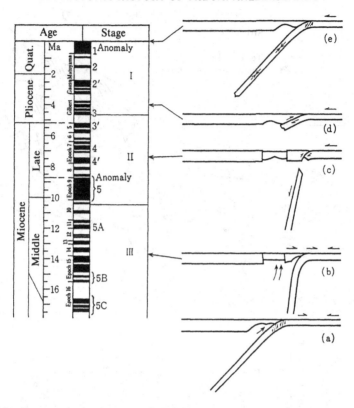

Fig. 10.9. Geotectonic development of the Northeast Japan Arc and evolution of the subducting slab (after NIITSUMA, 1979).

Japan, is represented by wavy deformation and faults oblique or perpendicular to that of the Nankai Trough. On the Pacific side, areas of upward and downward warping have occurred, depressions such as the Kii and Bungo Straits and mountains on the Kii, Muroto and Ashizuri Peninsulas have been formed. On the other hand, in the inner areas the block movements associated with a number of faults have taken place (HUZITA and KISHIMOTO, 1971; HUZITA and OTA, 1977; YOKOTA et al., 1978). As a result, the tilted mountain blocks and intermontane basins of the Kinki and Tokai districts were formed (Figs. 10.10(a) and (b)).

In the block movement areas, it is clear that the N-S trending faults are mostly of the reverse type; the NE-SW faults are right-lateral and the NW-SE faults are left-lateral. This suggests that in Quaternary time this region lay in a stress-field where the maximum compressional axis lay in a WNW-ESE direction (HUZITA, 1976). On the other hand, in Early Pliocene times (about

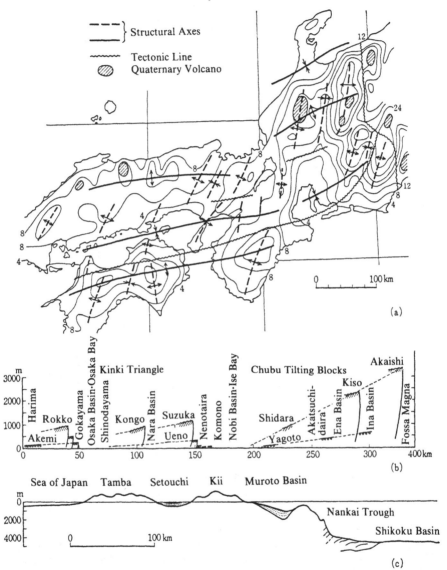

Fig. 10.10.   Up- and down-warping and block movements in Southwest Japan (a and b after
HUZITA and OTA, 1977; c after HUZITA, 1978). (a) Uplifting and subsidence axes in two
directions. (Contours for the summit level are in 400 m interval). (b) Schematic E-W section
showing tilted block mountains. Broken line: Summit level of lifted peneplane, Solid bar with
small circle: Higher terrace. (c) Schematic N-S section from Sea of Japan of Shikoku Basin.
Aspect ratio is 10:1. Strippled: Cenozoic strata.

5 Ma) the maximum compressional axis lay in a direction perpendicular to the axis of the Nankai Trough. This is partly suggested, by the direction of dikes intruded in the relevant areas.

The Miocene was a relatively quiet time in Southwest Japan. The rate of crustal movement was on the order of a few tens or hundreds times less than that experienced by the area in the late Quaternary. As in the Early Pliocene, the Middle Miocene stress field in the inner part of Southwest Japan, had its maximum compressional axis lying in a NNW direction, at right-angles to the Nankai Trough (KOBAYASHI, 1979). These conditions can be traced back to Early Miocene in the Shimanto belt. On the Pacific side, the strata, including those of Lower Miocene age, have been strongly folded, with axes parallel to the Nankai Trough.

The Miocene crustal movements in Southwest Japan were in general fairly weak. This may suggest that no or little subduction was taking place at the Nankai Trough at that time. However, the Nankai Trough formed a lateral displacement boundary between Southwest Japan and the Shikoku Basin, which was spreading in that period (30–15 Ma) (KOBAYASHI and NAKADA, 1978). But the volcanic activities in 14 Ma of the Setouchi and San-in-Hokuriku areas are not well explained by this idea, and much debate has taken place concerning Miocene plate tectonics of Southwest Japan.

The E-W compression of Quaternary times in Southwest Japan, is considered to have been caused by subduction of the Pacific plate in the Japan Trench, because the direction of compression is nearly the same with that in Northeast Japan (HUZITA, 1976; OKADA and ANDO, 1979). On the other hand, the deformational trends parallel to the Nankai Trough reflect subduction of the Philippine Sea plate at the Nankai Trough, which commenced in early (2–3 Ma) or late Quaternary (0.2–0.3 Ma).

10.10   Areas of Arc-Arc Junction in Central Hokkaido and Southern Fossa Magna

In the two areas, central Hokkaido and southern Fossa Magna, the eastern part of Hokkaido and the Izu peninsula are colliding with respect to the Honshu arc, resulting in compressional tectonics (DEN and HOTTA, 1973; OKADA, 1979; DICKINSON, 1978; MATSUDA, 1978). Neogene strata of both areas are marine, and mostly consist of turbidites. They include a number of coarse conglomerate beds, containing pebbles derived from contiguous mountain regions. Neogene formations of these areas, together with those of Paleogene, are strongly affected by compressional deformation (Fig. 10.8) and show imbricate structures, in which even the Quaternary sediments are involved.

The collision in central Hokkaido occurred in the Miocene (Fig. 10.11)

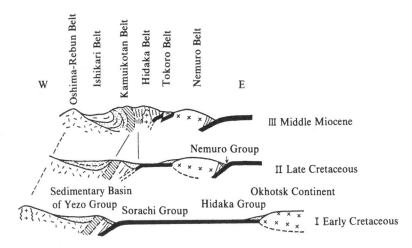

Fig. 10.11. Geotectonic development of Hokkaido (after OKADA, 1979).

(OKADA, 1979). As a result, the Hidaka Group terrane at the western margin of the Okhotsk Continent, was uplifted (the formation of the Hidaka Range), and the area to the west of the range subsided. In the latter area, thick marine turbidite beds were deposited and deformed. The Hidaka Range and the Shiranuka Hills are convex to the west, and accompanied by subsiding zones of the Ishikari Lowland to the west and the Tokachi Plain to the east. The structures are considered to have been formed by the westward projection of the outer zone of the Kurile Arc (KAIZUKA, 1975).

Fossa Magna is a large deformation zone, separating Northeast Japan from Southwest Japan. It is a region where the inner arc of Northeast Japan meridionally intersects the zonal structure of the older terranes of Southwest Japan. The western margin is limited by the Itoigawa-Shizuoka fault zone, along which the western older rock terranes are in contact with the eastern Tertiary to Quaternary terranes.

The zonal structure of Southwest Japan is bent acutely and forms the reverse V-shaped syntaxis in central Japan. It is considered to have resulted from the anti-clockwise and eastward movement of Northeast Japan (KOBAYASHI, 1941).

The Neogene strata of southern Fossa Magna include abundant submarine volcanic rocks, which are the product of northeastern Japan arc volcanism. However the depositional and deformational trend of the Neogene obliquely crosses the volcanic front and is almost parallel to that of the Shimanto belt (Fig. 10.12). In the offshore area, south of southern Fossa Magna, a subduction zone of the Philippine Sea plate existed since Miocene

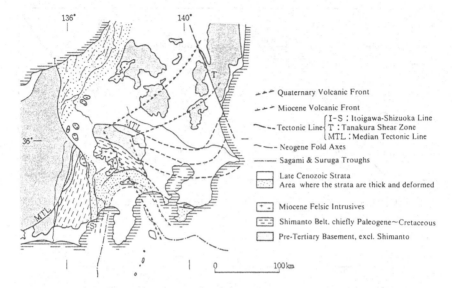

Fig. 10.12. Outline of the geologic structure of Fossa Magna and environs (data from geologic maps issued by Geological Survey of Japan).

times, this led to the collision of the Izu block in Quaternary time. The northward dipping thrusts, the radial arrangement of the maximum compressional axes, the exposure of crystalline schists in the Tanzawa Mountains and strong uplift of the Akaishi Mountains are all explained by the collision of the Izu block in the Quaternary (MATSUDA, 1978).

10.11 History of the Oceanic Plate Motion in the Vicinity of Japan

10.11.1 Before 40 Ma

Since the inception and spreading of the Pacific plate in the Early Mesozoic, the Kula Plate began subducting first and then (since 80 Ma) the Pacific plate at the northwestern margin of Pacific began its subduction (Fig. 10.13). Owing to the Pacific plate subduction in Hokkaido, the Okhotsk Massif migrated westwards and collided with western Hokkaido, at about 40 Ma (DICKINSON, 1978) or in Miocene (OKADA, 1979). The sedimentary terrane developed in that connection is the Ishikari-Kamuikotan belt.

On the other hand, the low P/T regional metamorphism and associated immense acid igneous activity took place in Southwest Japan in association with the subduction of the oceanic plate of the Shimanto terrane. According to UYEDA and MIYASHIRO (1973), the Kula-Pacific ridge was subducted underneath the Japanese Islands about 80 Ma ago, and caused these high

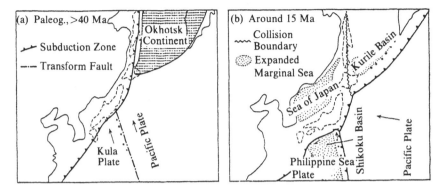

Fig. 10.13. Oceanic plates around the Japanese Islands (modified after DICKINSON, 1978; UYEDA and MIYASHIRO, 1973).

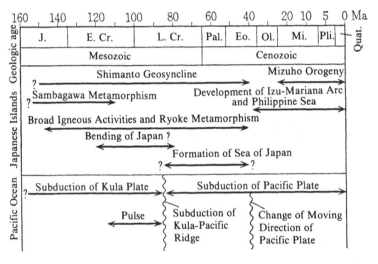

Fig. 10.14. Various geologic events in Japan in relation to plate development (after UYEDA and MIYASHIRO, 1973).

temperature phenomena. The opening of the Sea of Japan is also considered to have taken place in response to ridge subduction, though that occurred at a much younger period (Fig. 10.14).

A N-S trending transform fault existed to the south of the Japanese Islands in this period (HILDE *et al.*, 1979). Oceanic plates on both sides of the fault were subducting under the Japanese Islands. Aseismic ridges are inferred to have developed along the transform fault, which caused the northward bending of the zonal structure in central Honshu by the buoyant subduction (MATSUDA, 1978). The eastward migration, and anti-clockwise rotation of Northeast Japan, the NNW-trending left-lateral faults in northeast Japan (OTSUKI and EHIRO, 1978) and the supposed retroduction in Southwest Japan at the Median Tectonic Line (TAIRA, 1979) are the main events of this period.

10.11.2 About 40 Ma

As suggested by the bending of the Emperor-Hawaiian Seamounts chain, the direction of movement of the Pacific plate, changed from NNW to WNW at about 40 Ma. As a result, the site of the Japan Trench became an active subduction zone, and the N-S trending transform fault also changed into a subduction zone, along which the Izu-Ogasawara arc was formed. The present-day connections of the Kurile, Northeast Japan and Izu-Ogasawara arcs came into existence at that time.

10.11.3 After 40 Ma

The inner volcanic arc of the Izu-Ogasawara arc, formed by subduction of the Pacific plate since 40 Ma, was quickly split into two parts, both parallel to the volcanic front, and a new ocean appeared between them (KARIG, 1971). This is the Shikoku Basin, which was enlarged east- and westwards by the spreading axis, perpendicular to the Nankai Trough in the 30 Ma to 15 Ma period (KOBAYASHI and NAKADA, 1978). The newly formed eastern half of the split arc became the present day Izu-Ogasawara Arc, and the western half the Kyushu-Palau Ridge.

On the other hand, the subduction of the Philippine Sea plate at the Nankai Trough would have been weakened or ceased in the Neogene (later than 15 Ma). But in the eastern part of the trough, the subduction would have continued during Miocene times, resulting in the collision of the Izu terrane (the present-day Izu Peninsula) with Honshu in the Quaternary. Furthermore, in the central to western part of the Nankai Trough, the subduction recurred in the Quaternary (SUGIMURA, 1972; KANAMORI, 1977; MOORE and KARIG, 1976).

There is a triple junction, of the Eurasian (or North American), Philippine Sea and Pacific plates off central Honshu. It seems to be unstable, judging from the interrelationship of the meeting plates. Some discussions regarding this unstable triple junction have been made:

1) It has migrated southwards along the Japan Trench (MCKENZIE and MORGAN, 1969),

2) It has moved eastwards owing to the spreading (KARIG, 1971) of the Shikoku Basin or to the retreat of the subduction zone of the Pacific plate (KANAMORI, 1975),

3) It has remained at its present day position due to the northward shifting (KAIZUKA, 1975) or the widening (FITCH, 1972) of the Izu-Ogasawara Arc on the Philippine Sea plate, and

4) It is apparently stable, due to tectonic erosion of Northeast Japan due to the subducting Pacific plate (MATSUBARA, 1980).

REFERENCES

ADACHI, M., Paleogeographic aspects of the Japanese Paleozoic-Mesozoic geosyncline, *J. Earth Sci.*, *Nagoya Univ.*, **23/24**, 13–55, 1976.

ADACHI, M., The evolution of the Japanese Paleozoic-Mesozoic geosyncline, *The Basement of the Japanese Islands (Prof. H. Kano Volume)*, 119–141, 1979.

DEN, N. and HOTTA, H., Seismic refraction and reflection evidence supporting plate tectonics in Hokkaido, *Papers Meteorol. Geophys.*, **24**, 31–54, 1973.

DICKINSON, W. R., Plate tectonic evolution of North Pacific Rim, *J. Phys. Earth*, **26**, 1–19, 1978.

FITCH, T. J., Plate convergence, transcurrent faults, and internal deformation adjacent to southeast Asia and the western Pacific, *J. Geophys. Res.*, **77**, 4432–4460, 1972.

HADA, S., SUZUKI, T., YOSHIKURA, S., and TSUCHIYA, N., The Kurosegawa Tectonic Zone in Shikoku and tectonic environment of the Outer Zone of Southwest Japan, *The Basement of the Japanese Islands (Prof. H. Kano Volume)*, 341–368, 1979.

HILDE, T. W. C., UYEDA, S., and KROENKE, L., Evolution of the western Pacific and its margin, *Tectonophys.*, **38**, 145–165, 1977.

HORIKOSHI, E., Orogenic belts and plates in the Japanese Islands, *Kagaku (Science)*, **42**, 665–673, 1972.

HUZITA, K., The Quaternary tectonic stress states of Southwest Japan, *J. Geosci.*, *Osaka City Univ.*, **20**, 93–103, 1976.

HUZITA, K., Crustal movements and sea-level changes since Miocene in Southwest Japan in relation to the sedimentation and topographic surfaces, *Prof. N. Ikebe Volume*, 169–185, 1978.

HUZITA, K. and KISHIMOTO, Y., Neotectonics and seismic activity of the Kinki province, *Kagaku (Science)*, **42**, 422–430, 1972.

HUZITA, K. and OTA, Y., Quaternary tectonics. The Quaternary Period: Recent Studies in Japan, *Japan Association for Quaternary Research*, 127–152, 1977.

ICHIKAWA, K., MATSUMOTO, T., and IWASAKI, M., The evolution of the Japanese Islands, *Kagaku (Science)*, **42**, 181–191, 1972.

ICHIKAWA, K., MURAKAMI, N., HASE, A., and WADATSUMI, K., Late Mesozoic igneous activity in the inner side of Southwest Japan, *Pacific Geol.*, **1**, 97–118, 1968.

IGO, H., Re-examination of stratigraphy and geologic structure based on the conodonts in the eastern Mino terrain, *Prof. M. Kanuma Volume*, 103–113, 1979.

ISHIWADA, Y., IKEBE, Y., OGAWA, K., and ONITSUKA, T., A consideration on the scheme of sedimentary basins of Northeast Japan, *Prof. K. Huzioka Volume*, 1–7, 1977.

ISHIWATARI, A., Igneous petrogenesis of the Yakuno Ophiolite (Japan) in the context of the diversity of ophiolites, *Contrib. Mineral. Petrol.*, **89**, 155–167, 1985.

IWASAKI, M., Inferred basement rocks of the Sanbagawa terrane, *The Basement of the Japanese Islands (Prof. H. Kano Volume)*, 281–298, 1979.

KAIZUKA, S., Distribution of Quaternary fold, especially rate and axis direction in Japan, *Geogr. Rept., Tokyo Metropol. Univ.*, **3**, 1–9, 1968.

KAIZUKA, S., A tectonic model for the morphology of arc-trench systems, especially for the echelon ridges and mid-arc faults, *Japan. J. Geol. Geogr.*, **45**, 9–28, 1975.

KANAMORI, H., Seismic and aseismic slip along subduction zones and their tectonic implications, *Island Arcs, Deep Sea Trenches and Back-arc Basins* (M. Talwani and W. C. Pitman, III, eds.), Amer. Geophys. Union, 163–174, 1977.

KARIG, D. E., Origin and development of marginal basins in the Western Pacific, *J. Geophys. Res.*, **76**, 2542–2561, 1971.

KIMINAMI, K., TAKAHASHI, K., and MANIWA, K., The Cretaceous System in Hokkaido—Yezo and Nemuro Groups, *Assoc. Geol. Collabor. Japan Monogr.*, **21**, 111–126, 1978.

KIMURA, T., On the geologic structure of t Paleozoic group in Chugoku, West Japan, *Sci. Papers, Coll. Gen. Educ., Univ. Tokyo*, **10**, 109–124, 1960.

KIMURA, T., Structural development of Japan and plate tectonics, *J. Geogr.*, **86**, 54–67, 1977.

KITAMURA, N., Tertiary orogenesis in northeast Honshu, Japan, *Contrib. Inst. Geol. Paleontol., Tohoku Univ.*, **49**, 1–89, 1959.

KOBAYASHI, K., Subduction and uplifting of island-arc, *Earth Monthly*, **1**, 845–853, 1979.

KOBAYASHI, K. and NAKADA, M., Magnetic anomalies and tectonic evolution of the Shikoku inter-arc basin, *J. Phys. Earth*, **26**, 391–402, 1978.

KOBAYASHI, T., The Sakawa orogenic cycle and its bearing on the origin of the Japanese Islands, *J. Fac. Sci., Univ. Tokyo, Sec. II*, **5**, 219–578, 1941.

KOMATSU, M., MIYASHITA, S., MAEDA, J., OSANAI, Y., and TOYOSHIMA, T., Disclosing of a deepest section of continental-type crust up-thrust as the final event of collision of arcs in Hokkaido, north Japan, *Accretion Tectonics in the Circum-Pacific Regions* (M. Hashimoto and S. Uyeda, eds.), 149–165, 1983.

MATSUBARA, Y., Izu Peninsula and Philippine Sea Plate, *Earth Monthly*, **2**, 157–163, 1980.

MATSUDA, T., Collision of the Izu-Bonin arc with central Honshu: Cenozoic tectonics of the Fossa Magna, Japan, *J. Phys. Earth*, **26**, 409–421, 1978.

MATSUDA, T., NAKAMURA, K., SUGIMURA, A., Late cenozoic orogeny in Japan, *Tectonophys.*, **4**, 349–366, 1967.

MATSUMOTO, T., Timing of geological event in the circum-Pacific region, *Can. J. Earth Sci.*, **14**, 551–561, 1977.

MARUYAMA, S., BANNO, S., MATSUDA, T., and NAKAJIMA, T., Kurosegawa zone and its bearing on the development of the Japanese Islands, *Tectonophys.*, **110**, 47–60, 1984.

MCKENZIE, D. P. and MORGAN, W. J., Evolution of triple junctions, *Nature*, **224**, 125–133, 1969.

MIYASHIRO, A., Orogeny, regional metamorphism, and magmatism in the Japanese Islands, *Medd. fra Dansk Geol. Forening*, **17**, 290–446, 1967.

MOORE, J. C. and KARIG, D. E., Sedimentology, structural geology, and tectonics of the Shikoku subduction zone, Southwest Japan, *Geol. Soc. Amer. Bull.*, **87**, 1259–1268, 1976.

NAKAMURA, K. and UYEDA, S., Stress gradient in arc-back arc regions and plate subduction, *J. Geophys. Res.*, **85**, 6419–6428, 1980.

NIITSUMA, N., Geotectonic development of Northeast Japan Arc, *Kagaku (Science)*, **49**, 36–43, 1979.

OKADA, A. and ANDO, M., Active faults and earthquakes in Japan, *Kagaku (Science)*, **49**, 158–169, 1979.

OKADA, H., Geology and plate tectonics of Hokkaido, *Earth Monthly*, **1**, 869–877, 1979.

OKUDA, Y., KUMAGAI, M., and TAMAKI, K., Tectonic development of the continental slope and its peripheral area off Southwest Japan in relation to sedimentary sequences in sedimentary basins, *J. Japan. Assoc. Petrol Tech.*, **44**, 279–290, 1979.

OTSUKI, K. and EHIRO, M., Major strike-slip faults and their bearing on spreading in the Japan Sea, *J. Phys. Earth*, **26**, 379–389, 1978.

RESEARCH GROUP FOR ACTIVE FAULTS, Active Faults in Japan; sheet maps and inventories, Univ. Tokyo Press, 363 p., 1980.

SAITO, Y. and HASHIMOTO, M., South Kitakami Region: An allochthonous terrane in Japan, *J. Geophys. Res.*, **87**, 3691–3696, 1982.

SUGIMURA, A., Plate boundaries in the vicinity of Japan, *Kagaku (Science)*, **42**, 192–202, 1972.

SUGIMURA, A., MATSUDA, T., CHINZEI, K., and NAKAMURA, K., Quantitative distribution of late Cenozoic volcanic materials in Japan, *Bull. Volcanol.*, **26**, 125–140, 1963.

SUZUKI, T. and HADA, S., Cretaceous tectonic melange of the Shimanto belt in Shikoku, Japan, *J. Geol. Soc. Japan*, **85**, 467–479, 1979.

SUZUKI, T., HADA, S., and YOSHIKURA, S., Geotectonic development model of the Outer Zone of Southwest Japan, *Earth Monthly*, **1**, 57–62, 1979.

TAIRA, A., KATTO, J., and TASHIRO, M., Geologic history and tectonics of arc-trench system of Southwest Japan, *Chishitsu News*, No. 296, 27–40, 1979.

UYEDA, S. and MIYASHIRO, A., Plate tectonics and the Japanese Islands, *Geol. Soc. Amer. Bull.*, **85**, 1159–1170, 1974.

YOKOTA, S., MATSUOKA, K., and YASHIKI, M., Problems on the Cenozoic sedimentary blankets in the Shigaraki-Yamato plateau and its adjacent areas, central Kinki, Southwest Japan, *Chikyu Kagaku (Earth Science)*, **32**, 132–150, 1978.

BOOKS AND MAPS ON THE GEOLOGY OF JAPAN

Books

A few books in English on the geology of Japan have been published, though most of these are currently out of print.

1) T. Kobayashi (1941). The Sakawa orogenic cycle and its bearing on the origin of the Japanese Islands. *Journal of Faculty of Science, University of Tokyo, Section II*, Part 5, pp. 219–578.

   A long paper on geotectonic synthesis, based on classical geosynclinal theory. Noteworthy in having pointed out first, the asymmetrical zonal structure of the Japanese pre-Cenozoic geology, particularly in Southwest Japan.

2) F. Takai, T. Matsumoto, and R. Toriyama (eds.) (1963). *Geology of Japan*. University of Tokyo Press.

   A stratigraphic treatise, describing most of the sedimentary formations. Igneous and metamorphic rocks are rather briefly dealt with.

3) M. Minato, M. Gorai, and M. Hunahashi (eds.) (1965). *The Geologic Development of the Japanese Islands*. Tsukiji-Shokan Co., Tokyo.

   A synthesis also mainly based on geosynclinal theory. The orogenesis in Late Permian to Early Triassic times, called the Honshu orogenic movement, and the development of the Neogene Green Tuff region are particularly emphasized.

4) T. Yoshida (ed.) (1975). *An Outline of the Geology of Japan*. Geological Survey of Japan.

   A short book aptly describing the outline of the geology of Japan.

5) K. Tanaka and T. Nozawa (eds.) (1977). *Geology and Mineral Resources of Japan* (3rd edition), Volume 1. Geology. Geological Survey of Japan.

   A comprehensive treatise of the Geology of Japan. The descriptions are based principally on stratigraphical classification. Published as the explanatory text for the geological map of Japan on 1:1,000,000 scale. The map has been republished separately as an atlas.

6) A regional geology series in seven volumes has been published by the Asakura Shoten Publishing Company, Tokyo, but, all the volumes are written in Japanese.

243

Maps and Charts

a. *Geological maps*

A great number of coloured geological maps are issued by Japanese national and provincial authorities. They are usually have an appendix giving an explanatory text with an English summary. Names of geological units and localities are also given in English. Of these a few examples are mentioned below.

1) 1:2,000,000 scale in three sheets. Geological Survey of Japan.
2) 1:1,000,000 scale: An atlas together with a few maps covering various aspects. Geological Survey of Japan.
3) 1:500,000 scale: Covering most regions, except a part of Hokkaido and Nansei-shoto. Geological Survey of Japan.
4) 1:500,000 scale maps of the surface geology of Japan in six sheets. Published by Department of Economic Planning.
5) 1:200,000 scale: 44 sheets have been published to date by the Geological Survey of Japan. Most provincial offices also publish coloured geological maps of their administrative districts on this or similar scale.
6) 1:75,000 scale: 83 sheets were published mostly before and just after the Second World War. At present this series has been discontinued.
7) 1:50,000 scale: The standard geological map series now being published by Geological Survey of Japan. Usually accompanied by an explanatory text with English resume.
8) 1:2,000,000 scale compiled map series issued by the Geological Survey of Japan. Comprises maps of coal fields, oil fields, mines, hydrogeology, hot springs, volcanoes, active faults, hydrothermal districts, in addition to the metamorphic, metallogenetic and radiometric age maps. The latter three have English versions.

b. *Submarine charts and maps*

Various kinds of charts and maps of the submarine topography and geology are published by Hydrographic Office, Geological Survey and Geographical Survey Institute.

Charts published by Hydrographic Office

1) Charts of the neighbouring sea areas, on a scale of 1:8,000,000 in one sheet; and on a scale of 1:3,000,000 in four sheets.
2) Charts of the open oceanic region on a 1:1,000,000 scale, in 38 sheets.
3) Sets of four maps showing submarine topography, submarine geological structure, magnetism and gravity anomalies of the continental shelf areas on a 1:200,000 scale. About 80 sets have appeared.
4) Sets of two maps showing submarine topography and geological structure on a 1:50,000 scale of the near-shore regions. 71 sets have appeared.
5) Similar sets of 1:10,000 scale maps of the near-shore regions. 15 sets of the Soya strait and western off-shore Tsushima areas.

Maps published by Geological Survey

Maps showing the submarine geology of the sea floor sediments on a 1:200,000 scale include several near-shore regions, such as Koshikijima, Tsushima-Goto, Sea of Sagami, Kii strait and Ryukyu.

Maps published by Geographical Survey Institute

The submarine topographic, and environmental-condition maps at a 1:25,000 scale published in a two sheet-set of areas such as Toyohashi, Yokkaichi, Matsuzaka, Toba and so on. 18 areas have been covered by now.

INDEX